# Safety, Reliability, and Human Factors in Robotic Systems

# Safety, Reliability, and Human Factors in Robotic Systems

Edited by
James H. Graham, Ph.D., P.E.
University of Louisville, Louisville, Kentucky

VAN NOSTRAND REINHOLD
New York

Library of Congress Catalog Card Number 91-24584
ISBN 0-442-00280-7



Manufactured in the United States of America

Published by Van Nostrand Reinhold
115 Fifth Avenue
New York, New York 10003

Chapman and Hall
2-6 Boundary Row
London, SE1 8HN, England

Thomas Nelson Australia
102 Dodds Street
South Melbourne 3205
Victoria, Australia

Nelson Canada
1120 Birchmount Road
Scarborough, Ontario M1K 5G4, Canada

16 15 14 13 12 11 10 9 8 7 6 5 4 3 2 1

**Library of Congress Cataloging-in-Publication Data**

Safety, reliability, and human factors in robotic systems/edited by
James H. Graham.
p. cm.
Includes bibliographical references and index.
ISBN 0-442-00280-7
1. Robots. I. Graham, James H.
TJ211.S32 1991
629.8'92—dc20 91-24584
CIP

# Contributors

**Bangalore Amarnath,** Department of Industrial Engineering, University of Louisville, Louisville, KY

**Klaus M. Blache,** Advanced Engineering Staff, General Motors Corporation, Warren, MI

**James W. Collins,** Division of Safety Research, National Institute for Occupational Safety and Health, Morgantown, WV

**John R. Etherton,** Division of Safety Research, National Institute for Occupational Safety and Health, Morgantown, WV

**James H. Graham,** Department of Engineering Mathematics and Computer Science, University of Louisville, Louisville, KY

**Ramanchandra Kannan,** Department of Engineering Mathematics and Computer Science, University of Louisville, Louisville, KY

**Waldemar Karwowski,** Center for Industrial Ergonomics, University of Louisville, Louisville, KY

**Heikki Koivo,** Department of Electrical Engineering, Tampere University of Technology, Tampere, Finland

**Risto Kuivanen,** Occupational Safety Engineering Laboratory, Technical Research Center of Finland, Tampere, Finland

**Timo Malm,** Occupational Safety Engineering Laboratory, Technical Research Center of Finland, Tampere, Finland

**John E. McInroy,** Center for Intelligent Robotic Systems for Space Exploration, Rensselaer Polytechnic Institute, Troy, NY

**Hamid R. Parsaei,** Department of Industrial Engineering, University of Louisville, Louisville, KY

**James A. Peyton,** Manager of Standards Development, Robotic Industries Association, Ann Arbor, MI

**Mansour Rahimi,** Institute of Safety and Systems Management, University of Southern California, Los Angeles, CA

**George N. Saridis,** Center for Intelligent Robotic Systems for Space Exploration, Rensselaer Polytechnic Institute, Troy, NY

**Porter E. Smith,** Department of Engineering Mathematics and Computer Science, University of Louisville, Louisville, KY

**Robert H. Sturges, Jr.,** Department of Mechanical Engineering, Carnegie-Mellon University, Pittsburgh, PA

**Jyrki Suominen,** Occupational Safety Engineering Laboratory, Technical Research Center of Finland, Tampere, Finland

# Contents

# Preface

Today, we seem to be at the threshold of implementing robots with many of the characteristics long treated in a speculative fashion by science fiction writers and futurists. The initial arena of application has been in manufacturing situations to do rather simple tasks like material handling, spot welding, and spray painting. Careful attention to safety planning has been required because, although these industrial robots present many of the same hazards as conventional machine tools, they also have an element of flexibility and uniqueness not seen in other manufacturing equipment.

Today, many new applications for robotic systems in security, food preparation, laboratory operations, and human services are under development. As these new applications areas emerge, the level of interaction between robots and humans will inevitably increase, and safety may well become the critical consideration. This book attempts to provide both an overview of current robot safety issues and some insight into the multidisciplinary research and development that is currently being carried out to address the future needs in this essential area.

By conscious design, the chapters in this book represent a variety of topics relating to the unifying theme of robot safety. Robot safety is both an immediate practical concern and a long-range research effort, and both points of view are represented in the chapters that follow. It would be impossible to treat all of these topics exhaustively in a single volume, and so each contributor has attempted to summarize the key work in the area and to provide adequate references for further investigation by the reader. In particular, this book is oriented toward issues of designing safety and reliability in robotic systems. Hopefully, readers from a variety of backgrounds will benefit from the different chapters of this book and will be better prepared to address the robot safety and reliability problems in their own domains of interest.

The first chapter presents an overview of the issues involved in robot safety, reliability, and human factors, and lays the groundwork for succeeding chapters. Chapter 2 discusses the current state of industrial robot safety as determined through the efforts of the National Institute for Occupational Safety and Health. Chapter 3 discusses safety practices for industrial robot installations, and chapter 4 details the standards for robot safety as developed in the ANSI/RIA robot safety standard. Chapter 5 discusses reliability and safety considerations for an important subclass of robotic automation, the teleoperated robot. Chapters 6 and 7 discuss attempts to provide intelligent sensing and decision making capabilities for robotic systems. Chapter 8 discusses a study of the human factors involved in the interaction of humans and robotic systems. Finally, Chapter 9 provides a look at advanced research aimed at improving the reliability and safety of the next generation of intelligent automation.

I would like to thank, first and foremost, the contributors for their timely cooperation in preparing initial and revised drafts of their manuscripts. I would like to thank the editorial staff at Van Nostrand Reinhold, in particular Bob Esposito, Stephen Zollo, and Cynthia Savaglio, for their helpful suggestions and professional preparation of the final product. I would like to thank my professorial colleagues at the University of Louisville, Rensselaer Polytechnic Institute, and other institutions for their professional comments and suggestions on the subject matter to be covered in this book. I would also like to thank my colleagues in industry for their insight into current practices and problems in robot safety. Finally, I would like to thank the secretarial staff of the Speed Scientific School at the University of Louisville for their help in preparing the documents and correspondence involved in this project, in particular, Mary Mills, Gina Payne-Yunker, and Susan Cunningham.

# Safety, Reliability, and Human Factors in Robotic Systems

# 1

# Overview of Robot Safety, Reliability, and Human Factors Issues

James H. Graham

## 1. INTRODUCTION

The term "robot safety" may be interpreted in a variety of ways, the two most likely meanings being that one is concerned with either (a) preventing damage to the robot itself or (b) preventing the robot from damaging its environment, particularly the human component of that environment. While task (a) is praiseworthy considering the expense of modern industrial robots, it must in the long run remain subordinate to the latter activity, especially in regard to human safety.

In providing for human safety, most present robot installations rely on a combination of preventive maintenance for the robot, extensive operator training in safety procedures, and the installation of physical barriers, such as woven-wire fences, to keep humans out of the robot workspace. While these steps are reasonable and necessary for safe robot operation, there are many safety situations that they do not fully address. In particular, human personnel are required to be physically close to the robot in many aspects of robot set-up, programming, program debugging, and maintenance for the present generation of industrial robots. The use of mobile robots for both industrial and domestic applications is projected to increase. Finally, a key growth area is anticipated in the human services industries (Englehardt, 1985), where the very nature of the robot task is to interact closely with the human client.

It is the opinion of this author, based on several years of study, that the most reasonable and productive approach to the problems encompassed by this area lies in considering both the robot and the humans with which it must interact as a combined system. Thus, it is necessary both to consider enhancements to the robot (for example, expansion and improvements in sensing repertoire, sensory processing and integration algorithms, control, etc.) and to develop a better understanding of the human factors that govern the ways in which the human perceives and interacts with the robot.

Although both of the previously mentioned disciplines are in their infancy, the general questions of robot safety are beginning to get some attention, so that at least issues are being raised and examined. This chapter is intended as a partial survey of some of these issues and the corresponding research that is being undertaken to address these issues.

## 2. ROBOT FACTORS IN ROBOT SAFETY

There are a number of safety factors that are inherently influenced by the design and implementation of the robot and its related support equipment. Among these factors are the proper design of barriers, interlocks, and warning devices; design for the enhancement of the physical integrity and reliability of the robot hardware; sensory system development; and software reliability enhancement.

Robot safety equipment presently in use can roughly be classified into two groups according to function: (1) equipment that acts to prevent humans from coming into the workplace of an activated robot and (2) equipment used to detect and avoid humans within the robot workspace. Equipment in class 1 include physical barriers, such as fences and partitions, and also devices such as photoelectric light-beam curtains or capacitive fields that are interlocked to the robot controller in such a manner that crossing into the robot workspace disables the robot (Thompson, 1985).

In considering equipment from class 2, it is generally acknowledged that the best way to improve the robot side of the safety equation is to equip the robot sufficiently so that it can sense obstacles (human or otherwise) in its environment. Unfortunately, most present generation industrial robots have very limited sensory capabilities, and primarily repeat, or attempt to repeat, a predetermined sequence of trajectories. They have, at best, limited capability to detect obstacles in their paths, let alone, to distinguish inanimate obstacles from human beings. Obviously, substantial work must be done in the domains of robot sensing and control to remedy this situation!

The first reported work in this area in the United States was done by Kilmer and his colleagues (Kilmer, 1982) at the National Institute of Standards and Technology (NIST) [formerly National Bureau of Standards (NBS)]. The NIST researchers defined three levels of safety protection as follows: Level 1—perimeter-penetration detection around the workstation; Level 2—intruder detection within the workstation; Level 3—intruder detection very near the robot. Level 1 protection can include physical barriers as well as perimeter-monitoring devices such as photoelectric fences. Levels 2 and 3 protection require presence-sensing devices mounted either on or near the robot. These two levels are distinguished primarily by the response times required to avoid a collision. Level

3 is often characterized as a thin envelope around the robot itself, whereas Level 2 is within the reach of the robot but not in the immediate path of movement. The NIST researchers investigated the use of safety mats for Level 2 and ultrasound sensors mounted on the arm for Level 3 protection.

Research at Rennselaer Polytechnic Institute (RPI) has expanded upon the efforts at NIST and has concentrated on sensory protection for Levels 2 and 3. Four sensing technologies—ultrasound, infrared, microwave, and capacitive—were chosen as most closely matching the requirements for this task. The ultrasound and capacitive sensors were mounted on the robot for Level 3 protection, whereas the microwave and infrared sensors were mounted around the reachable robot workspace to give Level 2 protection. These sensors were tested on a Cincinnati Milacron T3 robot with good results in the laboratory setting (Graham and Meagher, 1985; Graham et al., 1986a; Meagher et al., 1983). Subsequent work at RPI has concentrated on refinement of these sensors and investigation of additional sensing modes (Millard, 1987).

Researchers at West Virginia University and the National Institute of Occupational Safety and Health have jointly investigated a combination of light curtains, pressure mats, and ultrasonics for safeguarding robot workstations (Sneckenberger et al., 1987; Tian and Sneckenberger, 1988). Researchers at the Technical Research Center of Finland working jointly with researchers at Tampere University have implemented a multisensory system of contact mats and light curtains to safeguard a welding robot (Kuivanen, 1988). Bennekers and Ramirez have reported the use of a video camera and a three-dimensional laser scanner for obstacle avoidance and safety in an aerospace assembly robot workstation (Bennekers and Ramirez, 1986).

The use of multiple sensors, as shown in several studies, has an advantage in that it can potentially overcome a defective or disabled sensor. However, the integration of disparate sensory data has proven to be a challenge because each sensor type has its own region of coverage, its own sensing characteristics, and its own inherent reliability. Recent work by this author has shown how a modified form of the Dempster–Shafer theory can be used for this purpose (Graham 1986b; Graham and Smith, 1988b). This approach has worked well in simulations because it uses a common mathematical formulation as the framework for each sensor type. Currently, the author is looking at ways to speed the computation and at methodologies for integrating the results with conventional robot workcell control architectures.

## 3. ROBOT SYSTEM RELIABILITY

Reliability is defined as the probability that an item will perform a required function under stated conditions for a stated period of time (Dhillon, 1983). Obviously, industrial robots that fail to perform properly, due to either partial or total functional failure, over extended periods of time would not satisfy the required economics for implementation in an industrial application, and robot system manufacturers strive to make robots as reliable as possible. On the other hand, industrial robots contain many elaborate components, and the manufacturer cannot be sure that the robot will be used within specified limits (Gelders and Pintelon, 1988).

Robots and other complex systems are often assumed to have relatively constant failure rates during most of their useful lifetimes. The hazard function for such a system is then a constant expressed as a number of expected failures per hour or, more commonly, the reciprocal is used to express mean time between failures (MTTR). An important related property is the availability of the robot, which may be defined as the probability that the robot is operational at some specified time. Robot system availability may be shown to be equal to the ratio of MTTF to MTTF plus mean time to repair (MTTR). Thus, reducing MTTR improves systems availability.

Preventive maintenance, per the manufacturer's recommendations, is the best insurance for reliable robot operation. A variety of operations research approaches can be used for optimal scheduling of preventive maintenance (Gelders and Pintelon, 1988). Once a failure has occurred, the primary consideration is rapidly to identify and correct the problem. Recent research has indicated the usefulness of using expert systems to speed up the diagnostic part of the repair process (Alexander et al., 1989; Graham and Alexander, 1990).

A final reliability-oriented factor that merits further study is the development of reliable robot system software. Existing techniques from the field of software engineering would appear to be appropriate (Shooman, 1983), but more investigation is needed to identify the specific software error modes that most directly impact robot safety. Recently, the issues of real-time software safety have received increased attention within the computer science field (Leveson, 1986; Parnas et al., 1990). Rahimi and his colleagues (Rahimi, 1988) have investigated the area of software safety in automation systems.

## 4. HUMAN FACTORS ISSUES IN ROBOT SAFETY

There are many ergonomic or human factors considerations that should be reviewed and incorporated into a properly designed robotic system. Many of these factors, such as the layout of robot control panels and teach-pendants (Levosinski, 1984; Podogorski and Boleslawski, 1990), training for personnel who operate or service the robot (Carrico, 1985), the design of safety barriers and interlocks (Bellino, 1985), etc., have a direct and obvious bearing on safety. In addition, human factors considerations appear in a number of areas of automation design which at first might not seem to require such considerations.

For example, a human factors evaluation of a robotic workstation is necessary if one considers the times when a human operator must be in the workstation (for maintenance, programming, etc.) (Helander, 1984). Several good surveys of human factors considerations that impact the safety of industrial robot installations have been presented recently (Noro and Okada, 1983; Parsons, 1986; Salvendy, 1986).

An important consideration is the speed at which a robot should be allowed to run when a human is in the workstation, for example, during teach-mode programming. The American National Standards Institute (ANSI, 1986) robot safety standard distinguishes between automatic mode and teach mode and specifies a maximum slow speed of 10 inches per second for teach-mode operations. Initial work by Nagamachi and his colleagues (Nagamachi and Anayama, 1983) and recent work by Karkowski and his colleagues have attempted to quantify the perception of maximum safe speeds by human subjects (Karwowski et al., 1987; Primovic and Karwowski, 1987).

Another important human factors area concerns how humans should be alerted to potentially dangerous situations involving robots (Stanton and Booth, 1990). Although the use of warning signs, audible alarms, and flashing lights is discussed in many of the previously cited studies, there does not seem to be a consensus on exactly where and how they should be deployed.

Finally, the areas of teleoperation and laboratory robotics have inherent human factors components associated within their scope. In teleoperation, a human attempts to guide a robotic device through some set of desired configurations using a high-bandwidth interface, such as a six-degree-of-freedom joystick or similar device. The human is an immediate and integral part of the control loop, and several studies of the resulting synergism and potential safety problems and limitations have been conducted (Malone et al., 1988; Ntuen and Park, 1988). Likewise in many laboratory robotics applications, humans and robots are working in close proximity, without the elaborate safeguards used for industrial installations.

## 5. SYSTEM ENGINEERING ISSUES IN ROBOT SAFETY

Many of the issues raised in the previous two sections pertain both to the robot and to the human operator, and are thus illustrative of the importance of a systems engineering approach as espoused by many investigators (Carrico, 1985; Graham and Meagher, 1985; Rahimi, 1986). It is critically important that standard safety systems methods be used to analyze the potential hazards and to plan for proper defensive measures, including operator education, for each robot installation.

An important systems task is the analysis of accident data to determine the proximate cause and any contributing causes, with a view to how future accidents can be avoided. A recent paper by researchers at Auburn University (Jiang and Gainer, 1987) discusses a cause-and-effect analysis of several U.S. robot accidents. Sugimoto and Kawaguchi present a fault-tree-analysis method for robot accidents and apply it to several Japanese robot accidents (Sugimoto and Kawaguchi, 1983).

A final systems task is the development of adequate and appropriate standards for the safe operation of robots. Most of the western industrialized countries have announced at least tentative standards. In the United States, this task was handled by a technical committee of the Robot Industries Association (RIA) and the result was endorsed and distributed by the American National Standards Institute (ANSI) as Standard R15.06 (ANSI, 1986). Input for the formulation of this standard was solicited from robot manufacturers, robot users, and the academic research community. This document sets standards relating to the construction, installation, maintenance, and use of industrial robot systems. Among its stipulations is a requirement that every robot have a hardware-based emergency stop that shall "override all other robot controls, remove drive power from the robot actuators, and cause all moving parts to stop." All operator stations, including teach-pendants, are required to have a clearly and uniquely identified emergency stop device. Following an emergency stop, restarting the robot shall require an operator-initiated restart sequence performed outside the restricted work envelope. Furthermore, the robot is required to be "designed and constructed so that any single, reasonably forseeable failure will not cause hazardous motion of the robot." The installation section provides that the robot shall be installed in accordance with the robot manufacturer's specifications and in such a way to avoid interference with other structures and machines. The restricted work envelope of the robot is required to be "conspicuously identified, for example, by means of signs, barriers, line markings on the floor, railings or the equivalent."

Operators of the installed robot system shall be trained in the proper operation of the control actuators and safeguards of the robot system, and in how to recognize, and to react to, the known hazards of the system. Normal operating personnel will be prohibited from entering the restricted work envelope of the robot by one or more safeguarding devices, including barriers, interlocked barriers, perimeter-guarding devices, presence-sensing safeguarding devices, and awareness devices. Specific personnel, such as programmers and maintenance workers, will receive special training in the performance of their tasks, must perform a set of specified checks before entering the workspace while power to the robot is on, and are required to leave the workplace before initiating automatic motion of the robot. During programming, the robot is restricted to the low-speed setting, which is defined by the standard to a maximum of 250 millimeters per second (10 inches per second).

## 6. CONCLUSIONS

This chapter has attempted to summarize briefly the major areas of research currently under investigation in the development of safe and reliable robots. These areas include human factors issues, the design of the robot itself, systems engineering issues, and standards development. Although the issues are becoming more clearly defined, and certain standards have been established for current industrial robots, the conclusion in many aspects of this subject is that much more research must be performed, in preparation for future and more sophisticated robot applications.

The following chapters provide snapshots both of the present state of robotic safety practices and of the directions presently being developed and evaluated for future generations of robots. Taken together, it is hoped that these selections will provide insight into the technical nature of this subject. Chapter 2 discusses the current state of robot safety as determined by the investigations of the National Institute of Occupational Safety and Health. Chapter 3 discusses in detail the safety practices used in industrial robot installations in a major manufacturing application. Chapter 4 presents a more detailed analysis of the RIA/ANSI robot safety standard. Chapter 5 discusses safety and reliability of teleoperated robot systems. Chapters 6 and 7 discuss systems for advanced sensing and decision making to achieve enhanced safety in robots. Chapter 8 discusses the interaction of robots and humans from a human factors perspective, and Chapter 9 looks at future designs for safe and reliable intelligent machines.

Although this volume cannot hope to provide final answers to all of the complex safety, reliability, and human factors issues that confront designers and users of today's robotic systems, it is hoped that it will help point the direction for the development of enhanced safety features in the future generations of robotic systems, so that these systems may come closer to fulfilling their potential in improving the quality of human life.

### References

ANSI, 1986. *American National Standard for Industrial Robots and Robot Systems—Safety Requirements,* ANSI/RIA R15.06-1986. New York: American National Standards Institute.

Alexander, S., Graham, J. and Vaidya, C. 1989. Issues in the diagnosis of manufacturing systems. In *Proceedings of AAAI Symposium on Artificial Intelligence in Manufacturing.*

Asimov, I. 1984. *Machines that Think.* New York; Holt, Rinehart, and Winston.

Bellino, J. P. 1985. Design for safeguarding. In *Robot Safety,* eds. M. C. Bonney and Y. F. Yong. Berlin: Springer Verlag.

Bennekers, B. and Ramirez, C. 1986. Robot obstacle avoidance using a camera and a 3-D laser scanner. In *SME Vision 86.* pp. 2.57–2.65. Dearborn, MI: SME.

Carrico, L. R. 1985. Training and design for safe implementation of industrial robots. In *Robot Safety,* eds. M. C. Bonney and Y. F. Yong. Berlin: Springer Verlag.

Dhillon, B. S. 1983. *Reliability Engineering in Systems Design and Operation*. New York: Van Nostrand Reinhold.

Englehardt, K. G. 1985. Applications of robots to health and human services, SME Technical Paper MS85-0587, Dearborn, MI: Society of Manufacturing Engineers.

Etherton, J. 1988. Unexpected motion hazard exposures on a large robot assembly system. In *Ergonomics of Hybrid Automated Systems,* ed. W. Karwowski. New York: Elsevier.

Etherton, J. Safe maintenance guide for robotic workstations. Washington, DC: NIOSH.

Etherton, J. and Sneckenberger, J. 1990. A robot safety experiment varying robot speed and contrast with human decision cost. *Applied Ergonomics,* 21(3):231–236.

Gelders, L. F. and Pintelon, L. M. 1988. Reliability and maintenance. In *International Encyclopedia of Robotics*. ed. R. Dorf. New York: Wiley.

Graham, J. and Meagher, J. 1985. A sensory-based robotic safety system. *IEE Proceedings* 134 (Pt. D):183–189.

Graham, J. et al. 1986a. A safety and collision avoidance system for industrial robots. *IEEE Transactions on Industry Applications* IA-22:195–203.

Graham, J. 1986b. An evidential approach to robot sensory fusion. In *IEEE Intl. Conf. Systems, Man and Cybernetics*. pp. 492–497. New York: IEEE.

Graham, J. and Millard, D. 1987. Toward development of inherently safe robots. *SME Robots 11*. pp. 9.11–9.22, Dearborn, MI: SME.

Graham, J. 1988a. Overview of research issues in robot safety. In *Ergonomics of Hybrid Automated Systems*. ed. W. Karwowski. New York: Elsevier.

Graham, J. and Smith, P. 1988b. Computational considerations for robot sensory integration. In *Proceeding of Symposium on Advanced Manufacturing*. pp. 115–118.

Graham, J. and Alexander, S. 1990. Computer based diagnosis for integrated manufacturing systems. *Proceedings of Second International Conference on Computer Integrated Manufacturing*. pp. 669–673.

Helander, M. 1984. Safety design of robot workplaces. In *Human Factors in Organizational Design and Management,* ed. H. Hendrick. Amsterdam: Elsevier.

Jiang, B. and Gainer, C. 1987. A cause and effect analysis of robot accidents, *Journal of Occupational Accidents* 9:27–45.

Karwowski, W. et al. 1987. Human perception of the maximum safe speed of robot motions. Proc. Annual Meeting of Human Factors Society, New York.

Karwowski, W. et al. 1988. Human perception of the work envelop of an industrial robot. In *Ergonomics of Hybrid Automated Systems,* ed. W. Karwowski. New York: Elsevier.

Kilmer, R. D. 1982. Safety sensor systems for industrial robots. In *SME Robots 6 Conference*. pp. 479–491 Dearborn, MI: SME.

Kumekawa, S. and Sugimoto, N. 1988. AVG safety system designed for preventing hazardous human contact. In *Ergonomics of Hybrid Automated Systems,* ed. W. Karwowski. New York: Elsevier.

Kuivanen, R. 1988. Experiences from the use of an intelligent safety sensor with industrial robots. In *Ergonomics of Hybrid Automated Systems,* ed. W. Karwowski. New York: Elsevier.

Leveson, N. 1986. Software safety: why, what and how. *ACM Computing Surveys* 18:125–163.

Levosinski, G. J. 1984. Teach control pendants for robots. SME Tech. Paper MS84-0944. Dearborn, MI: SME.

Malone, T. et al. 1988. Operator issues in the control of telerobotic system. *SME Robots 12,* pp. 3.75–3.82. Dearborn, MI: SME.

Meagher, J. et al. 1983. Robot safety and collision avoidance. *Professional Safety* 28:14–18.

Meieram, H. 1988. Man-machine interface for mobile robotic systems. In *Ergonomics of Hybrid Automated Systems,* ed. W. Karwowski. New York: Elsevier.

Millard, D. 1987. Animate sensing for industrial automation safety. In *SME Robots 11* Conference. Chicago, April 1987. Dearborn, MI: SME.

Nagamichi, M. and Anayama, Y. 1983. An ergonomic study of the industrial robot. *Japanese Journal of Ergonomics,* 19:259–264.

Nagamichi, M. 1988. Ten fatal accidents due to robots in Japan. In *Ergonomics of Hybrid Automated Systems,* ed. W. Karwowski. New York: Elsevier.

Noro, K. and Okada, Y. 1983. Robotization and human factors. *Ergonomics* 26:985–1000.

Ntuen, C. and Park, E. Human factor issues in teleoperated systems. In *Ergonomics of Hybrid Automated Systems,* ed. W. Karwowski. New York: Elsevier.

Parnas, D. L., van Schouwen, J., and Kwan, S. P. 1990. Evaluation of safety-critical software. *Communications of the ACM.* 33:636–648.

Parsons, H. M. 1986. Human factors in industrial robot safety. *Journal of Occupational Accidents* 8:25–47.

Parsons, H. M. 1988. The future of human factors in robotics. *SME Robots 12,* pp. 3.99–3.113. Dearborn, MI: SME.

Podogorski, D. and Boleslawski, S. 1990. Ergonomics in the design of robot teach pendants. In *Ergonomics of Hybrid Automation Systems II,* eds. W. Karwowski and M. Rahimi. New York: Elsevier.

Primovic, J. and Karwowski, W. 1987. Effect of automation on safety performance. Proc. Ninth Intl. Conf. Production Research, Cincinnati, OH. pp. 2694–2700.

Rahimi, M. 1986. Systems safety for robots. *Journal of Occupational Accidents* 8:127–138.

Rahimi, M. 1988. Critical issues in the safety of software dominated automated systems. In *Ergonomics of Hybrid Automated Systems,* ed. W. Karwowski. New York: Elsevier.

Shooman, M. 1983. *Software Engineering: Design, Reliability and Management.* New York: McGraw-Hill.

Slavendy, G. 1986. Human factors in planning robotic systems. In *Handbook of Industrial Robotics,* ed. S. Y. Nof. New York: Wiley.

Sneckenberger, J., Kittiampton, K., and Collins, J. 1987. Interfacing safety sensors to industrial robotic workstations. *Sensors* 35–37.

Stanton, N. and Booth, R. 1990. Alarm initiated actions. In *Ergonomics of Hy-*

*brid Automation Systems II*, eds. W. Karwowski and M. Rahimi. New York: Elsevier.

Stauffer, R. 1989. Robotic safety: A goal, not a given. *Manufacturing Engineering*, 102:73–74.

Sugimoto, N. and Kawaguchi, K. 1983. Fault tree analysis of hazards created by robots. *Proceedings of 13th International Symposium on Industrial Robots*. pp. 9.13–9.28.

Thompson, C. 1985. Safety interlock systems. In *Robot Safety*, eds. M. C. Bonney and Y. F. Yong. Berlin: Springer Verlag.

Tian, J. and Sneckenberger, J. 1988. Performance evaluation of three pressure mats as robot workstation safety sensors. In *Ergonomics of Hybrid Automated Systems*, ed. W. Karwowski. New York: Elsevier.

# 2

# NIOSH Research to Prevent Injuries and Fatalities Associated with Industrial Robotics

James W. Collins and John R. Etherton

## 1. INTRODUCTION

NIOSH is the U.S. federal agency responsible for conducting research to prevent work-related injuries, illnesses, and fatalities. NIOSH's Division of Safety Research (DSR) is the Institute's focal point for research to prevent occupational injuries and fatalities. The mission of the DSR is to reduce the frequency and severity of injuries in the industrial workplace by developing and conducting research projects with real-world intervention potential. The objective of this chapter is to present an overview of the hazards of industrial robot operations and the DSR's research and response to combat these hazards.

The DSR has investigated two robot-related fatalities and has obtained reports on numerous robot-related injuries and near misses. In Japan, where more robots are currently used, ten fatalities have been documented (Nagamachi, 1988). In the United States, as more automation is introduced into the workplace and worker exposure increases, there is a serious potential for injury to workers who interact with robotic systems. The number of operators and maintenance personnel who will be subject to risks in robot work zones will grow significantly according to an Office of Technology Assessment (OTA) study of professions affected by automation (Office of Technology Assessment, 1984). Because of the potential for robot-related fatalities and the increasing exposure of workers to the hazards of industrial robots, the DSR formulated a robotics research program in 1983 to address a range of safety issues in the field of robotics. This program has developed comprehensive information to assist in the safe installation, use, maintenance, and supervision of industrial robot systems. The DSR developed this information by conducting intramural research on human interaction with robotic systems and sponsoring extramural research on the degree of worker exposure to robot hazards, robotics safety sensor development, and human performance at robotized workstations.

The NIOSH/DSR robotics research program represents a welcome departure from the traditional approach to occupational safety research. The DSR robotics safety research program was developed to deal with the emerging technology of robotics before large numbers of injuries occurred. Normally, research is conducted in areas where high risk is indicated by injury and fatality data. For example, the large number of amputation injuries to mechanical power press operators led to NIOSH research to reduce the risk of injury to press operators.

Automotive companies began developing robotic safety standards in 1980 when the installed industrial robot population in this country was approximately 1000. The Robotic Industries Association was then encouraged in 1983 to provide the leadership to develop a consensus standard for the safety of industrial robots. The DSR robotics research program has assisted in the introduction of voluntary robotics safety standards.

## 2. ROBOT INJURY AND FATALITY SURVEILLANCE

Although history has shown that the introduction of new technology is eventually followed by the development of safe performance criteria and standards, that interval can be plagued with worker injury and death. The DSR robotics research program is based in part on the fact that shortening the interval between the initial utilization of robots and the application of appropriate safeguards will decrease the risk of injury and death for workers exposed to associated hazards.

### 2–1. First U.S. Robot-Related Fatality

After being notified by the Consumer Product Safety Commission in the state of Michigan about the first robot-related fatality in the United States, the DSR sent a research team to the work site to conduct an investigation. The fatal injury occurred in a small die-casting plant with approximately 280 employees and 24 die-casting machines. During the 2 years prior to the incident, the company had automated two die-casting systems with robots. The robot involved in the incident had been purchased second-hand from the original owner, thus eliminating the assistance of the robot manufacturer in the installation of the robotic system. The workstation layout and safeguarding devices were designed and installed by in-house employees who had no previous experience with robotics. On the day of the incident, the male victim was overseeing a robotized die-casting workstation and inspecting finished castings. See Figure 2–1. The robot in this system was programmed to extract a casting from the die-cast machine, dip it into a quench tank, insert it into a trim press, and then repeat the cycle. The

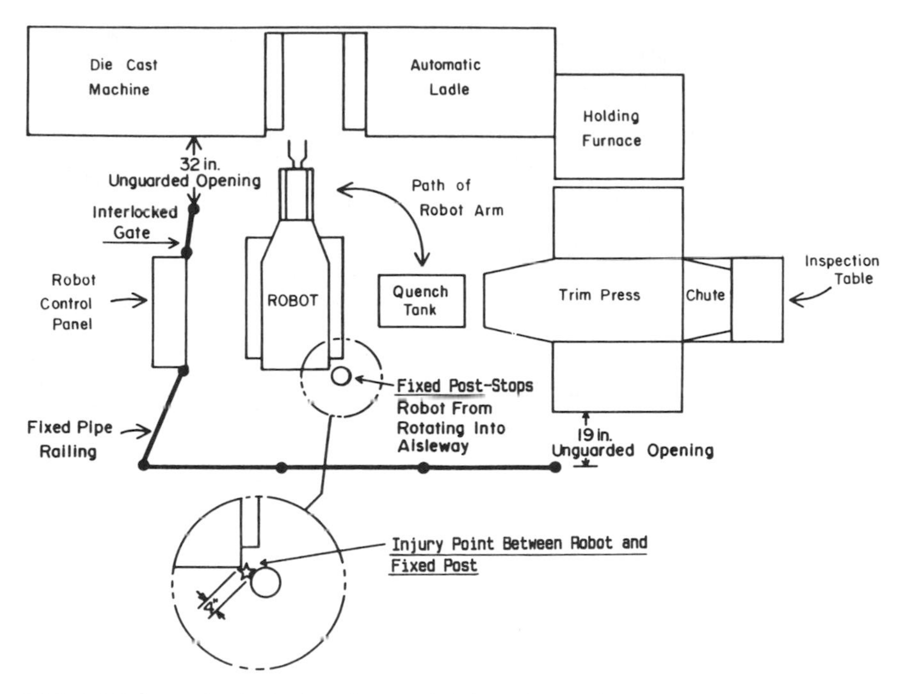


**FIGURE 2–1.** Overhead view of automated die-casting workstation.

primary safeguard at the time of the fatality was an interlocked gate in a partial perimeter safety railing, which had two unguarded openings that permitted undesired access to the robot workstation while the system was in operation. The victim could have entered the workstation of the robot through either of the two unguarded openings (32 in. wide and 19 in. wide) at opposite ends of the safety railing or climbed over, under, or through the railing.

It is not known why the worker was inside the safety railing while the robot was operating automatically, but an air gun used during debris removal was found inside the workstation. It is presumed that the die-cast operator entered the workstation to remove scrap metal that had accumulated on the floor from the trim press during his shift. Plant procedure specified stopping the robot before performing this clean up task. The only emergency stop button was located on the control panel outside of the enclosed workstation.

The victim was discovered by another employee who was manually operating an adjacent die-casting machine. The accident site could not be seen from the neighboring workstation. The other worker stated that after hearing a continual hissing noise for a period of 10–15 min, he went to investigate. The victim was found pinned in a slumped but upright position between the back of the robot arm

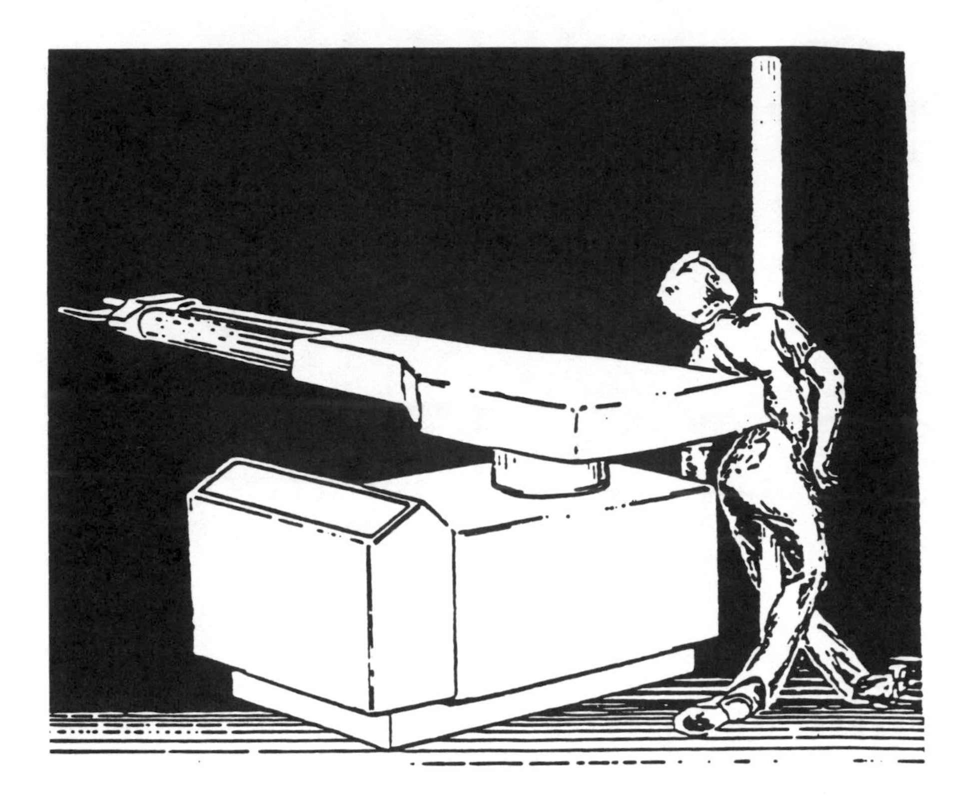

**FIGURE 2–2.** Artist's sketch of first robot-related fatality in the United States.

and a steel pole that was intended to restrict the range of robot arm motion (see Figure 2–2). The robot stalled when it contacted the man's body and continued to apply pressure on his chest for approximately 25 min. The victim died 5 days later in the hospital. The pathology of the injuries is unknown because the next-of-kin would not allow an autopsy. However, X-rays of bones and internal organs showed no signs of crushing injury.

## 2–2. Second U.S. Robot-Related Fatality

In the second U.S. robot-related fatality in 1986, an operator of an automated machining system was found with his back on a conveyor while a robot's end effector (end-of-arm tooling) applied downward pressure on his chest. The automated workstation utilized a robot to transfer starter motor frames into and out of three milling machines. See Figure 2–3. Each milling machine had a loading conveyor that carried starter housings up to the machining area and an

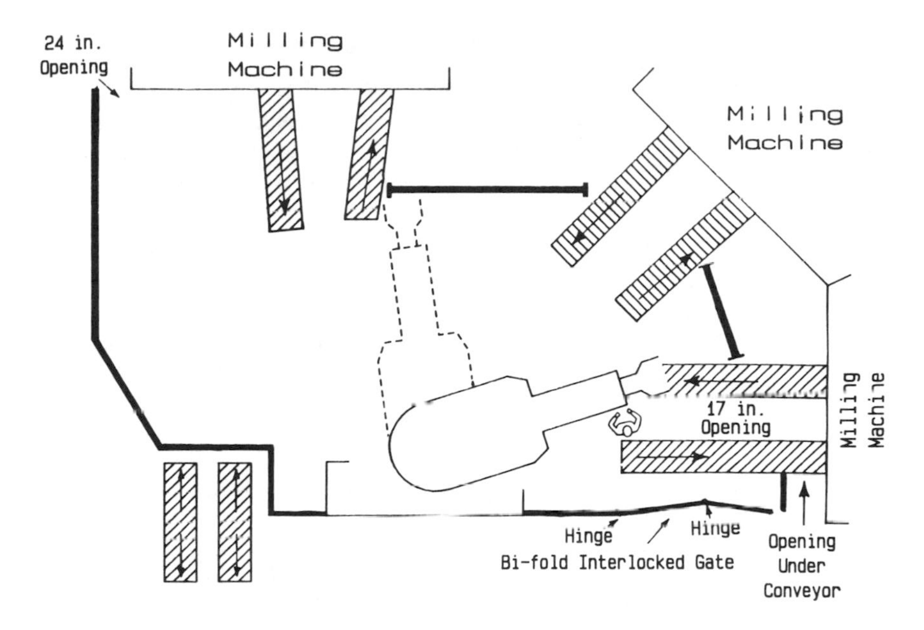


**FIGURE 2–3.** Overhead view of automated machining station.

unloading conveyor that returned finished housings where they could be picked up and transferred out of the work cell. The industrial robot was programmed to randomly load and unload a conveyor on demand. A safety fence with a bi-fold gate enclosed the workstation. The bi-fold gate was designed to allow for reconfiguration to permit safe manual operation of one milling machine while the other two operated automatically. The operator of this system had been involved with the initial installation of this automated workstation and had worked with it for 9 years.

The operator was found near death with his back pinned against the top of a loading conveyor by the robot. For some unknown reason the employee had entered the robot's work envelope while the robot was operating automatically. The robot arm was executing a normal programmed move when it hit the operator in the chest and knocked him down onto the loading conveyor. This pinch point occurred during the robot's normal operating sequence when the robot delivered or retrieved a starter housing from the conveyor. The robot's work envelope could be entered through either of two unguarded openings, which were not equipped with interlocked gates or safety devices (see Figure 2–3) and only half of the bi-fold gate was equipped with an interlock. In another incident 10 months prior to this fatality, this worker had been struck in the head by the same industrial robot.

### 2–3. Comparison of the Two Robot-Related Fatalities

A common factor in the two U.S. robot-related fatalities was that the victims had entered the safeguarded area of the robotic workstation during automatic operation. In both cases, the safeguarding was intended to prevent access to the robot's work zone. Opening an interlocked gate to enter the work zone would stop automatic operation of the robot. Although both workcells were equipped with interlocked gates, the safety perimeters had openings that allowed worker access to the robot's work zone without stopping automatic operation of the robot.

There was a substantial difference in robot system operation experience between the two victims. The fatally injured operator in the first incident had finished his initial training 3 weeks before the incident, whereas the operator involved in the second incident had 9 years of experience.

## 3. NIOSH/DSR RESEARCH TO PREVENT ROBOT-RELATED INJURIES

The following five subsections review NIOSH/DSR robot-related hazard exposure modeling and simulation experiments to quantify human performance when working near robots. This research was conducted to evaluate applicable sections of the "American National Standard for Industrial Robots and Robot Systems–Safety Requirements" (American National Standards Institute, 1986). The objective of this standard is to enhance the safety of personnel associated with industrial robot systems.

### 3–1. Modeling the Exposure of Workers to Industrial Robot Hazards

Technicians are occasionally required to work near moving industrial robots when performing maintenance tasks. These tasks place the technician at risk of being trapped between a moving part of the robot and a fixed object in the robot's work zone. A lockout/tagout procedure need not be followed in such cases according to the ANSI R15.06 "Safety Standard for Industrial Robots and Industrial Robot Systems-Safety Requirements." However, the ANSI standard does require that personnel performing maintenance or repair tasks shall have sole control of the robot and robot system while drive power is available to the system.

To compare robot-related injuries with the injury experience associated with other occupational hazards, exposure-based injury rates need to be computed. In order to do this, information on the amount of worker exposure to robot hazards needs to be collected (Etherton, 1987a, 1988a). The most obvious hazard is pinch points created by robot motion toward fixed objects, which can injure robotics technicians. Two ways that workers may become exposed to the hazard of robot motion toward fixed objects are as follows: (1) they enter the robot's work zone while the robot is operating automatically or (2) they enter the robot's work envelope during a maintenance task that must be done with drive power available. To assess the degree of hazard exposure among robot maintenance personnel, a project was carried out by a NIOSH/DSR researcher at a large manufacturing company (Etherton, 1988b).

This project identified potential hazard exposures for robotic maintenance workers. Data were collected on a 19-cell automated assembly line that included 25 SCARA robots dedicated to assembling a large number of small parts. The line has already been in operation for a long period of time and the maintenance personnel under study were regularly assigned to maintain this line. Examination of maintenance logs and discussions with maintenance personnel were used to identify and quantify potential exposures to hazards during maintenance activities. For each robot in this 25-robot assembly system, there was an average of 5.4 min per 8-hr workday when maintenance activities required personnel to enter the robot's work zone.

In this study, time of potential hazard exposure was defined as the time a robot was logged in a "down for maintenance" status (i.e., time that a maintenance person performed a troubleshooting, adjustment, or repair task). Table 2–1 shows the total mean exposure times in minutes and the total numbers of activities for troubleshooting, adjustment, and repair tasks. Also shown is the breakdown between inside workcell/power available activities, that is, activities inside the protective workcell perimeter with power available, and other mainte-

**TABLE 2–1 Average Total Time That Maintenance Workers Were Exposed to Robotic Hazards During 5-Month Study**

| | Other Maintenance Activities | | Inside Workcell/ Power Available Activities | | Total Activities | |
|---|---|---|---|---|---|---|
| Task Type | Mean time (min) | *n* | Mean time (min) | *n* | Mean time (min) | *n* |
| Trouble-shooting | 353 | 2 | 223 | 13 | 240 | 15 |
| Adjust-ment | 80 | 26 | 138 | 25 | 109 | 51 |
| Repair | 89 | 40 | 82 | 14 | 87 | 54 |
| Total | 93 | 68 | 144 | 52 | 115 | 120 |

nance activities. Examination of this table reveals that the troubleshooting task required the most time. The adjustment task required the next longest time, primarily for inside workcell/power available activities.

Repair took the least time, and required about equal times for both types of activities. For the maintenance actions studied, 76% were performed with power available to the robot and 43% required maintenance personnel to enter the robot's work envelope with power available. The data for the 5 months under evaluation indicated that the mean exposure time for troubleshooting jobs was twice as long as for repair and adjustment. The mean exposure time for inside workcell/power available activities was 144 min. The mean exposure time for inside workcell/power available activities was significantly longer than the other maintenance activities (144 min vs 93 min).

Furthermore, inside workcell/power available activities took more time overall, but only because the task distribution was different between the activities. The other maintenance activities were more heavily weighted toward repair, whereas the inside workcell/power available activities were weighted more toward the adjustment and troubleshooting tasks.

Based on the results of this study, the following suggestions were made to the company:

1. The company should investigate ways to ensure effective energy control while maintenance work is done inside a workcell's safety perimeter with power available to the robot, that is, for inside workcell/power available activities. Safety devices (e.g., ultrasonic, voice recognition, capacitance, pressure mats) that protect workers inside the safety perimeters should be considered.
2. The company might also use current maintenance hazard exposure as a baseline against which to compare contemplated changes in the system design. Design changes that can reduce maintenance exposure times would have a positive effect on hazard level.

## 3–2. Modeling Human Reactions to Robot Motion

Certain human/machine/environment system factors influence how well technicians react to moving robots. The ANSI safety standard states that reduced robot speed increases worker safety during programming and maintenance tasks and that a specific slow speed of 25 cm/sec should not be exceeded (American National Standards Institute, 1986). A systems approach was used to assess the speed of the robot end effector as a risk factor for the hazard of a technician being trapped by a robot (Etherton, 1987b). Leamon (1980) proposed a closed-loop human–machine model. (See Figure 2–4.) The model organizes system elements in a problem-oriented manner, which reflects actual industrial demands. The

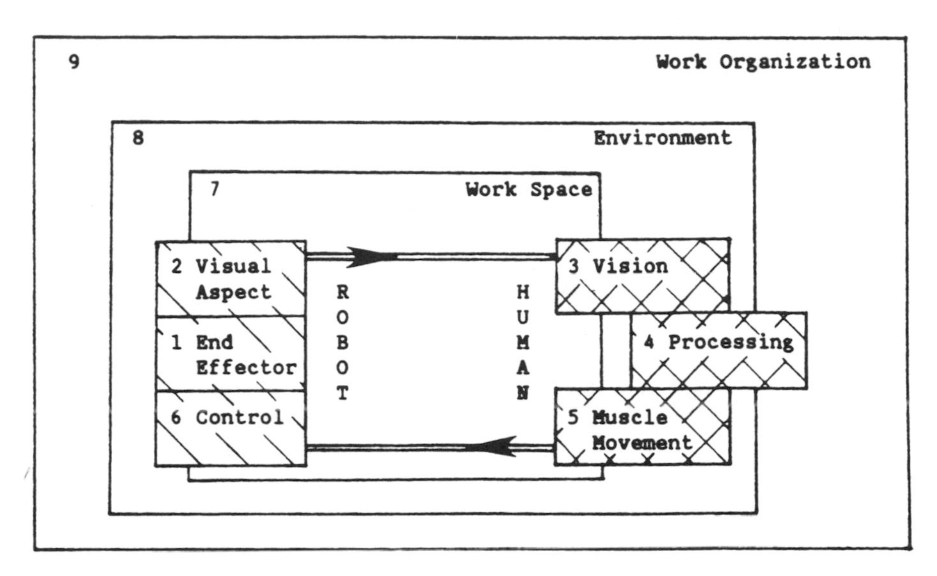


**FIGURE 2–4.** Human/machine/environment system model applicable to the NIOSH/DSR study of slow robot speed for safe maintenance. Modeled after Leamon (1980).

nine elements in Figure 2–4 are hypothesized to relate to human reaction to slow robot motion:

1. Robot tooling or end effector.
2. Visual aspect of the robot arm.
3. Human vision.
4. Human decision processing of visual image.
5. Human muscle movement.
6. Robot axis control.
7. Objects in the robot's work envelope.
8. Lighting and other environmental factors.
9. Training and established work procedures.

The NIOSH/DSR application of the Leamon model suggests where interactions may exist, but it does not indicate how much these interactions contribute to overrun distance. Overrun distance is defined as the distance that the robot arm travels between when a hazardous condition is recognized and the robot arm is stopped by manually depressing an emergency stop button. The model shows that both the human and machine elements operate in a dependent relationship. For a given robot speed this value can be translated into an overrun distance:

$$\text{Overrun distance} = \text{Velocity of robot arm} \times \text{Time required to stop robot.}$$

Further modeling aspects for predicting human reaction time to dangerous robot motion were developed by Helander and Karwan (1987) for NIOSH/DSR. Their model emphasizes two critical aspects of human reaction to unexpected slow robot motion: the reaction time and the probability that a reaction will take place. In the Helander–Karwan model, robot stopping time is a summation of the five components in the following formula:

$$\text{Robot stopping time} = \text{Human perception time} + \text{Human decision time} + \text{Human movement time} + \text{Robot controller time} + \text{Braking time.}$$

## 3–3. Workplace Observations for Designing Robot Safety Simulation Experiments

It was desired that NIOSH/DSR human factors/robotics laboratory experiments be designed to simulate actual workplace conditions. Site visits were first made to examine industrial robots that load and unload parts from a large hot forging press and to further evaluate conditions under which personnel work near robots that are moving at slow speed (Etherton, 1987c). During the site visit researchers (1) observed and discussed the robotic operations with technicians who work with the robots, (2) surveyed safety-related factors during maintenance at the workstation, and (3) took measurements at workstations on sound levels, lighting levels, and contrast between robots and background.

It was found that personnel in two job categories were potentially exposed to robot motion hazards at the facility: production operators and programmer/maintainers. The programmer/maintainer typically had 8–10 years of experience on robotic equipment. Maintenance records indicated there were about 7 hr/month of downtime for a two-robot system. It was estimated that about 1 hr of that downtime involves personnel being in the robot's work envelope with drive power available to the robot. The estimates of typical repair time ranged from 15 to 60 min, the frequency of maintenance activities ranged from three per week to one per day, and the use of emergency stops for system shutdown ranged from once every 2 weeks to once every 10 weeks. Each workstation had emergency stop capabilities at the gate, the press control, the robot control panel, and each side of the press frame where a robot was located.

During interviews with NIOSH/DSR researchers, robotic workers indicated that they did not feel there was any penalty associated with initiating an emergency stop, even though it took 30–60 min to restart a system that was operating correctly. Robotic workers said they had experienced from 1 to 25 instances of uncontrolled robot motion per year. Problems with servo valves resulting in one axis swinging out of control were the most frequent. The largest and most potentially hazardous forces were generated when the base axis swung out of control. Slow robot arm speed in this facility was slightly above 25 cm/sec.

## 3–4. Experiments on Safe Slow Speeds for Robots

The speed of a robot is a safety factor during teaching or any other mode of operation. However, it is too simplistic to assume that the slower the mandated speed of a robot arm, the lower the risk. Considerations on what an appropriate value should be for slow speed include:

- The time required to move to avoid the robot arm.
- The speed should allow for efficient programming and thus be acceptable to those expected to use slow speed.
- Situations where because of the position of the teacher a slower speed is appropriate.
- Contrast between end effector and background.
- Decision cost in using an emergency stop.
- Slow speeds above the ANSI 25-cm/sec threshold.
- Lighting level.
- Frequency of unexpected robot motion.

An initial laboratory simulation experiment to evaluate 25 cm/sec as an appropriate safe slow speed for robots during programming was conducted for the NIOSH/DSR by Etherton et al. (1988). This experiment utilized a General Electric P50 industrial robot that was programmed to simulate unexpected dangerous movements. The time required for subjects to hit an emergency stop was measured for these unexpected dangerous movements. On certain movements, which were unknown to the subject, the robot was programmed to simulate a dangerous condition by overrunning a preset position. The probability of an overrun occurring was 0.10. Subjects viewed the robot from angles of 0°, 45°, and 90°, and the robot moved at one of four speeds (15, 25, 35, and 45 cm/sec). Three different age groups of human subjects (20–30, 31–40, and 41–60 years) were studied. Results of the experiment were that when the robot arm was moving at 25 cm/sec, the arm traveled a mean overrun distance of 7.77 cm and a maximum of 16 cm before the emergency button was depressed. At 45 cm/sec, a mean of 10.9 cm and a maximum of 20.5 cm overrun distance was observed. The simulation experiment showed that a subject with their hand near the emergency stop and no distractions achieved reasonable stopping distances at the current slow speed standard of 25 cm/sec.

The next NIOSH/DSR experiment to evaluate human response to various slow speeds (Etherton and Sneckenberger, 1990) primarily examined the decision cost of hitting the emergency stop when the robot is operating correctly and halting production unnecessarily. A simulated work environment was created in which subjects were to make decisions to stop the robot. The goal for all subjects was to minimize downtime. The decision cost for false alarm and for missed hazard (missed signal) was expressed in terms of minutes of lost worktime (downtime). Three levels of false alarm cost were used. (See Table 2–2.) The goal for all subjects was to minimize downtime. The experiment monitor provided subjects with verbal feedback on their downtime score. In this experiment the cost of a missed hazard was not trivial, but neither was it so large as to become the dominant criterion. It was felt that this is also the case in the industrial workplace, because all hazardous conditions may not result in immediate injury.

Significant differences in how close the robot arm could get to a subject who was attempting to stop dangerous robot motion were found for robot arm speed, decision cost, and the interaction between arm speed and decision cost. At the fairly high level of ambient light (52 footcandles) used in this experiment, different levels of visual contrast did not have a significant effect on reaction time.

Results of this experiment suggest that programmers and maintenance personnel should not be reluctant to stop a slow moving robot. Reaction to slower robot arm speeds seems to include a decision-making component that increases reaction time. The results further indicate that robotics technicians will be at minimal risk when working near robots that are moving 25 cm/sec or less, provided their hand is near an emergency stop button. These results are limited to the degree that it is difficult to simulate all the conditions that contribute to the general task of working near a moving robot.

Beauchamp (1989) extended these NIOSH/DSR experiments by building a simulated robotics workstation in which he varied the level of overhead illumination and ambient noise level while test subjects performed simple and complex manual tasks when observing the robot move at different speeds. A hazardous situation was simulated when the robot made a move toward the subject. Results were measured in reaction time to the unexpected movement.

Significant effects were found for illumination (the National Safety Council recommended 500 lux was found adequate), task (divided attention caused longer response times), and robot speed. Beauchamp concluded from these

**TABLE 2–2 Levels of Cost (Equipment Downtime) in Relation to False Alarm and Missed Hazard**

| Regime | Cost of False Alarm (min) | Cost of Missed Hazard (min) |
|---|---|---|
| No cost | 0 | 60 |
| Medium cost | 10 | 60 |
| High cost | 60 | 60 |

experiments that the National Safety Council's recommended slow speed of 15 cm/sec would allow personnel time to reach for an emergency stop, whereas the ANSI's 25 cm/sec would not provide protection when longer reach times were involved. Suggestions were made to train workers to maintain attention when near the robot, to use adequate lighting (500–1000 lux), and to paint the robot a color which is highly contrasting with its background.

## 3–5. An Experiment on Locating Emergency Stop Buttons on Robot Teach-Pendants

Control pendants are hand-held portable control and display devices used for teaching and recording the movements of an industrial robot. A survey of control pendants revealed that there is no standardization of hand-held pendants for

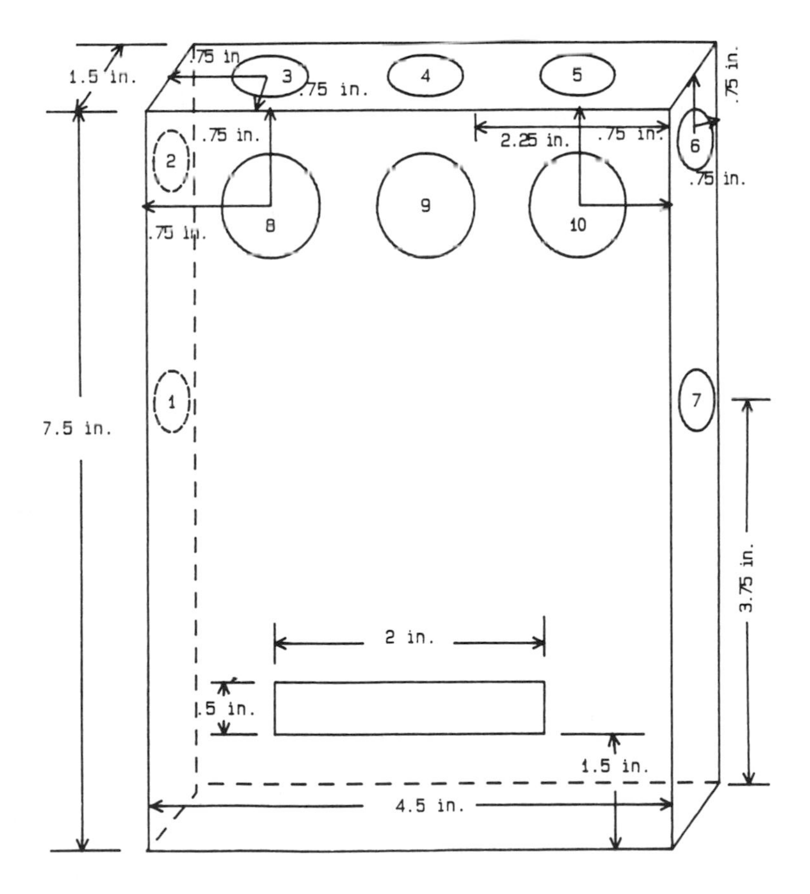


**FIGURE 2–5.** Front view of pendant simulator illustrating locations 1–10.

robotic applications. The question explored in a research project by Collins (1989) was: "In an emergency situation, when operator response time is critical, what effect do location and diameter of the emergency stop button have on the time required for a person to initiate an emergency stop?" To answer this question, a pendant simulator was designed and constructed to represent a typical single-handed control pendant that could accommodate emergency stop buttons of various diameters at 10 different locations (see Figure 2–5). Button diameters of 0.5 and 1 in. were tested at two locations on the left-side surface, three on the top surface, two on the right-side surface, and three on the front surface of the pendant simulator. The pendant simulator was hardwired to a digital timer, which recorded reach times to within 0.001 sec. A General Electric P50 floor-mounted five-axis industrial robot was programmed to provide the stimulus for the test subject to hit the emergency stop button.

Table 2–3 summarizes the mean and range values for hand reach time by diameter and location of emergency stop buttons. The insignificant variations between mean reach times for locations (1 and 2), (3, 4, and 5), (6 and 7), and

**TABLE 2–3 Mean and Range Reaction Times for All Subjects**

| Button Location | Range (sec) | | Number of misses | Twelve Subject Mean (sec) |
|---|---|---|---|---|
| | Low | High | | |
| | | Button Diameter = 0.5 in. | | |
| 1. | 0.120 | 0.300 | 1 | 0.164 |
| 2. | 0.117 | 0.289 | 1 | 0.169 |
| 3. | 0.124 | 0.219 | 1 | 0.163 |
| 4. | 0.122 | 0.319 | 0 | 0.167 |
| 5. | 0.114 | 0.314 | 1 | 0.167 |
| 6. | 0.110 | 0.208 | 3 | 0.144 |
| 7. | 0.097 | 0.275 | 1 | 0.147 |
| 8. | 0.087 | 0.391 | 1 | 0.138 |
| 9. | 0.086 | 0.194 | 1 | 0.120 |
| 10. | 0.086 | 0.257 | 1 | 0.127 |
| | | Button Diameter = 1 in. | | |
| 1. | 0.111 | 0.195 | 1 | 0.147 |
| 2. | 0.127 | 0.254 | 0 | 0.158 |
| 3. | 0.107 | 0.216 | 3 | 0.156 |
| 4. | 0.112 | 0.233 | 0 | 0.163 |
| 5. | 0.131 | 0.211 | 0 | 0.161 |
| 6. | 0.101 | 0.206 | 1 | 0.142 |
| 7. | 0.093 | 0.198 | 0 | 0.138 |
| 8. | 0.063 | 0.179 | 0 | 0.121 |
| 9. | 0.085 | 0.185 | 0 | 0.113 |
| 10. | 0.055 | 0.173 | 2 | 0.110 |

(8, 9, and 10) indicated that for the final analysis these locations could be combined and examined as surfaces. Not only was the average reach time for all subjects faster on the front surface of the pendant, but all subjects' individual average reach times were faster for the front surface. Table 2–3 also shows clearly that subjects could on the average reach the 1-in. button faster than the 0.5-in. button at all 10 locations.

Analysis of the study data revealed that human subjects were able to depress the 1-in.-diameter emergency stop button an average of over 7% quicker than the 0.5-in.-diameter button at all 10 locations. The fastest average reach time occurred on the front surface (0.128 sec) of the pendant simulator. Based on the analysis of the study data, emergency stop buttons larger than 0.5 in. in diameter on teach-pendants should be considered. Data from this study also revealed that depressing emergency stop buttons that only required unilateral hand motion (from the subject's starting hand position to depressing the emergency stop button) significantly reduced reach times.

## 4. NIOSH-SPONSORED RESEARCH ON ROBOT CONTROL SYSTEMS AND SAFETY SENSORS

NIOSH-sponsored research at West Virginia University led to the development of a three-level safety sensor system (Sneckenberger et al., 1987) and the testing of safety mats (Tian and Sneckenberger, 1988). In the three-level safety sensor system, a light curtain provided the first level of intruder detection at the workstation perimeter. If the sweeping beam of light produced by the curtain was broken by a solid object, the safety system responded by activating a yellow flashing light and reducing the robot's speed by 30%. Pressure mats were used on the floor area around the base of the robot. When pressure was applied to the mat, a flashing red light and an audible beeper were activated and the robot's speed was reduced by 60%. An ultrasonic sensor mounted on the robot's arm provided the third level of protection. This sensor emitted an ultrasonic signal in pulse form and the time required for the echo to be received was converted to a distance. If this distance was less than a precalibrated distance, a loud continuous horn was activated and the robot manipulator was halted. Further testing of the pressure-sensitive mats revealed that these safeguarding devices are highly reliable in a laboratory environment.

NIOSH-supported research at Rennselaer Polytechnic Institute (Millard, 1989) has led to the development of a capacitance-based safety sensor that mounts near a robot's manipulator. This sensor provides an adjustable volume of protection by creating a capacitance field that travels with the robot's manipulator. If a person intrudes into this capacitance field, drive power is removed and robot motion ceases.

## 5. GUIDELINES TO PREVENT ROBOT-RELATED INJURIES

NIOSH researchers have developed general guidelines pertaining to the following four areas: workstation design, training, management participation, and safe maintenance guidelines (Etherton and Collins, 1990; NIOSH, 1984; Sanderson et al., 1986).

### 5–1. Workstation Design

The American National Standards Institute (ANSI) Safety Requirements for Industrial Robots and Robot Systems, R15.06, developed by the Robotic Industries Association (RIA) (American National Standards Institute, 1986), should be followed in designing, operating, and managing a safe human–robot system. The standard states: "The means and degree of safeguarding should correspond directly to the type and level of hazard presented by the robot application. Safeguarding may include, but is not limited to, safeguarding devices, barriers, interlock barriers, perimeter guarding, awareness barriers and awareness signals." The following design guidelines can be followed to assist in the safe installation of robotic workstations:

**1.** Because industrial robots generally require more space than humans, placing robots in existing production lines may create pinch-point hazards. To minimize pinch points, space utilization problems should be carefully considered during the design of robotic workstations.

**2.** Manual controls that initiate automatic operation of the robot should be located outside the perimeter safeguarding. This will require employees to be outside the work envelope before initiating automatic operation of the robot and other equipment in the workstation. Consider using multiple E-stops, which are accessible within the workspace.

**3.** Workstation designers should be knowledgeable of the servicing requirements of the robot so that foreseeable scheduled and corrective maintenance can be performed easily and safely.

**4.** The installation of steel posts or any other fixed object near robot work envelopes that are not necessary to the robot operation violates a basic principle of safe engineering design and does not comply with the provision of ANSI/RIA 15.06. Although intended to eliminate the hazard created by erratic movement of the robot arm, the first U.S. robot-related fatality clearly demonstrates that fixed objects can create lethal person-sized pinch points.

**5.** In both of the U.S. robot-related fatalities, there were openings in the barrier system that a person could easily enter without opening the interlock. The

ANSI/RIA robotics safety standard states that "An interlocked barrier shall prevent access to the restricted work envelope except by opening an interlocked gate." Neither of the installations where the U.S. fatalities occurred met this criterion. If an interlocked barrier is intended, there should be no openings through the barrier except through the interlocked gate. In spray painting or other applications where toxic vapors could be emitted, a nonporous but rigid enclosure with adequate ventilation can provide an adequate perimeter safeguard.

**6.** Safeguarding can be evaluated to include appropriate controls to be used for preventing robot motion that would cause injury and provide for quick recovery to productive operation. Where feasible, these controls should be passive, such as safety floor mats or light screens. Controls that are passive or easily used and provide for efficient recovery will reduce risks because they will be accepted and employees will not look for methods to bypass the safeguarding system.

## 5–2. Training

Safety training should be provided for all personnel who will be programming, operating, or maintaining industrial robot systems. For this training to be effective, it must pertain to the worker's job for a specific robot application. This safety training should familiarize the employee with all potentially hazardous energy sources of the robot, the full range of robot motion, potential pinch points, placement and function of emergency stop buttons, review of available safety standards, and the designated means of safeguarding. Training should emphasize the hazards associated with entering the robot's work envelope and safe stopping and restarting procedures.

Programmers should be trained to ensure that robots are operating at a slow speed during programming, to avoid creating unnecessary pinch points, and to avoid placing themselves in existing pinch points. Training is not only needed for inexperienced workers, but also for experienced programmers, operators, and maintenance workers. This training allows experienced workers to remain up-to-date with technological advances and operational changes, and reveals problems or hazards they have experienced. Management can build a safety program around comprehensive training, but should not limit safety efforts to training only.

## 5–3. Management Participation

Managers and supervisors should be trained on the safe operation and maintenance of robots in their factory so they can establish and enforce a robotics safety policy. The company should have a clearly written safety policy that has been explained to all personnel who will be working with robots. This safety policy should state which personnel are authorized to work with robots.

Safeguarding needs should be assessed before purchasing an industrial robot. Buyers should take advantage of available sources for advice on incorporating safeguarding systems. Information can be obtained from robot manufacturers, automated systems specialists, the Robotic Industries Association, and government agencies such as NIOSH and OSHA.

There is an important role for supervisors to play in preventing robot-related fatalities. The job description for a supervisor who is responsible for robot operators can include the following requirements:

**1.** Maintain equipment and workplace in a safe condition. This requires knowing what safeguards are to be used on robotic workstations, what makes them effective, how they work as a system, and what needs to be done if their effectiveness is compromised. Supervisors should work with those assigned to implement the safety system on robotic workcells, making them aware of shortcomings of elements in proposed or operational safety systems. In particular, if perimeter guarding is used to safeguard the robotic workstation, the barrier should be equipped with an interlocked gate and should completely enclose the work envelope. This includes ensuring that safety system designers adequately understand the actual tasks that can expose workers to robotic system hazards. Safety devices should fail to a safe state for single-component failure. In other words, failure of a component of the safety device should not result in any danger to personnel or damage to plant facilities or the product.

**2.** Provide instructions on the purpose and effectiveness of safeguards on the robot workstation. This should include instruction on the dangers of bypassing safeguards, how to test safeguards to ensure they are functional, and correcting unsafe practices such as operators entering the robot's range of motion with the robot operating automatically.

**3.** Establish a process to ensure that employees are informed of safe work procedures for the specific robots on which they work. The procedures, including deenergization and lockout/tagout procedures, should include identifying and avoiding pinch points, safe methods when two people use controls during the same maintenance action, effective use of emergency stop devices, and how to help a fellow worker who may become trapped in the workstation. The lockout procedure should follow OSHA requirements (OSHA, 1989). As an example, the lockout procedure used during a major repair on a robot installation would include: (1) shutting off power to the robot using the normal sequence; (2) securing power at hydraulic, electrical, and pneumatic sources; (3) locking and tagging these energy sources; (4) relieving sources of stored energy; and (5) verifying that the area is safe before beginning the repair and before reenergizing the system.

**4.** Assign qualified people to perform robot operation, maintenance, and programming jobs. Workers who are qualified by training or experience are less likely to make errors that could harm themselves or fellow workers or damage the workcell.

**5.** Supervise operators as they work. This means having someone check on workstation operators where one works alone. Be aware that operators who monitor automated tasks may take risks because they become bored, overconfident, or lose concern for safety. Supervisors should be familiar with the safeguarding system and prescribe work practices for robot workcells to monitor effectively the use of these safety systems. The supervisor should be able to convey to employees that safeguarding systems should be used as intended and that it is the responsibility of each employee to observe safe work practices and not to override safeguarding systems. Employees are normally dedicated to maintaining the continuous operation of manufacturing systems. If the safeguarding method interferes, they are motivated to bypass the safeguard. If the method of stopping the system is inconvenient to use or makes it difficult to recover to productive operations after using it, they will be motivated to perform their task without stopping the system. Supervisors must not ignore or tolerate unsafe practices. If the intended safeguarding method is determined to cause inconvenience or interfere with the intended task, the supervisor can obtain assistance from safety or controls engineers to find a more acceptable method of safeguarding.

## 5–4. Safe Maintenance Guidelines

In 1988, NIOSH published guidelines to help robot users develop systems and procedures for safe robot maintenance (Etherton, 1988c). NIOSH developed these guidelines to help industry prevent injuries as industrial robotics evolves in the American workplace. The most practical way to discuss the guidelines is to focus on the three following conclusions:

**1.** Effectively designed safety systems for robotized workstations should protect personnel who periodically enter the robot's work envelope. In many cases, prescribed human tasks are randomly performed near the robot. The safety system should protect a person who enters a robot's work zone from the potential hazards associated with the robot and other machinery in the workcell. Therefore, tasks that require personnel to work near a robot, especially in the robot's immediate range of motion while it is under power, should only be done after the robot and associated machinery are in a selected safe condition.

**2.** Because robotized systems may be complex, a systematic approach for analyzing human safety during robot maintenance is advised. Before implementing proposed safety measures, their effectiveness should be evaluated with regard to foreseeable human entry into the workstation. The operation of automated equipment and associated safety devices should be clearly explained to all personnel who will be interacting with the equipment. A systematic analysis for safe maintenance of robotic workstations should be performed by various qualified persons. For example, a person qualified in machine control could assess whether failure of a programmable control device could result in dangerous movement of electrical-, hydraulic-, or pneumatic-driven elements that still have power available during a maintenance task. Someone skilled in labor-management relations could promote the use of safety devices. Safety professionals could provide information on choices among protective devices and procedures, safety regulations, and safety engineering principles.

**3.** Evaluation of maintenance policies and records will help determine the degree of potential hazard exposures inside robot work zones. Evaluation of maintenance records can show the level of worker exposure to various robotic hazards.

In using these guidelines it is also helpful to remember that valuable information can be obtained from discussions with employees who maintain the robot system. These employees can be consulted to determine the tasks that they actually perform and to determine if the safeguarding devices and prescribed work practices are effective and convenient to use and follow. Seek employees' input to modify controls, devices, and work practices for improving convenience and efficiency, thus improving the probability of use and the resulting reduction of risks.

## 6. PRIORITIES FOR CONTINUED ROBOT SAFETY RESEARCH

A roundtable to assess priorities in machine safety research met June 5, 1989, in Cincinnati, OH, in conjunction with the Annual International Ergonomics and Safety Conference (Etherton and Myers, 1991). Research needs relating to robot safety were one of the considerations at this roundtable. Discussion led to the identification of two compelling research needs:

1. Robot-related injury surveillance: The establishment of a comprehensive database that compiles information on industrial-robot-related injuries is needed to guide future prevention research.
2. Expert systems for robotic safety sensors: This area was seen as most compelling because of the need in industry for off-the-shelf sensors to safeguard workers at industrial robot workstations.

## 7. DISCUSSION

Since its inception in 1983, the robot safety research program at NIOSH/DSR has had many successes. Unfortunate fatalities that have occurred have been responded to with investigations and publications focusing on factors which can be avoided in other robotic systems. Guidelines for safety during robot maintenance, the task with greatest exposure to robot hazards, have been published. The ANSI performance requirement on a safe slow speed for robots has been tested under realistic conditions and has thus far been found to be adequate. Empirically based suggestions on the size and location of emergency stop buttons on hand-held control pendants have been made to the appropriate consensus standards committee.

Potential exposure of people to the hazards of moving robots will increase as the robot population increases. Although reliability improvements in robot control components can be expected, the reliability of human elements (vision, decision, motor response) are likely to remain unchanged. Further study is needed in human reliability to define the limits of human performance in an industrial robot system and thereby indicate the level of safeguarding required to protect exposed workers.

Safety research on robots that load and unload industrial presses should be promoted to remove workers from the persistent injury and amputation problems involving power press operation. Research in this area will provide answers to the question of how many injuries can be prevented with robotized systems.

The factors of illumination, noise, contrast ratio, and attention to a secondary task have been evaluated at least once for human response to various robot slow speeds. There is a need to look at other stressors such as fatigue, stress, age, body position, and safety training.

Simulation experiments and industrial cooperation have been two of the principal themes of the NIOSH program in robotic safety. Experimental simulation of potentially hazardous conditions is a safe way to study human–robot system safety performance. Cooperating on research projects with labor, the robotics industry, academia, and other research organizations increases the prevention impact of NIOSH's robotic safety research. NIOSH researchers will continue to conduct research and work cooperatively with other members of the machine safety research interest community to reduce the injury and fatality toll among America's work force.

**References**

American National Standards Institute. 1986. *Safety Standard for Industrial Robots and Industrial Robot Systems,* R15.06–1986, New York: ANSI.

Beauchamp, Y. 1989. A study of human performance in the event of an unexpected robot motion. Unpublished Doctoral Dissertation, West Virginia University, 1989.

Collins, J. 1989. Experimental evaluation of emergency stop buttons on hand-held teach pendants. Proceedings of the Annual Meeting of the Human Factors Society.

Etherton, J. 1987a. Automated maintainability records and robot safety. In Proceedings of Annual Reliability and Maintainability Symposium, Philadelphia, pp. 135–140.

Etherton, J. 1987b. Systems considerations on robot end effector speed as a risk factor during robot maintenance. In Proceedings of the Eighth International System Safety Conference, New Orleans.

Etherton, J. 1987c. Site Visit Report, Division of Safety Research, 944 Chestnut Ridge Rd., Morgantown, WV.

Etherton, J. 1988a. Human factors in managing robot system downtime. In *Success Factors for Implementing Change: A Manufacturing Viewpoint,* ed. K. Blache. pp. 209–221. Dearborn, MI: SME.

Etherton, J. 1988b. Unexpected motion hazard exposures on a large robotic assembly system. In *Ergonomics of Hybrid Automated Systems,* ed. W. Karwowski, pp. 411–419. Amsterdam: Elsevier.

Etherton, J. 1988c. Safe maintenance guidelines for robot workstations. NIOSH Technical Report 88–108.

Etherton, J., Beauchamp, Y., Nunez, G., and Ahluwalia, R. 1988. Human response to unexpected robot movements at selected slow speeds. In *Ergonomics of Hybrid Automated Systems,* ed., W. Karwowski. Amsterdam: Elsevier.

Etherton, J. and Collins, J. 1990. Supervisor awareness may prevent robot-related fatalities. *Professional Safety* March.

Etherton, J. and Myers, M. 1990. Machine safety research at NIOSH: thinking about the future. *International Journal of Industrial Ergonomics* 6:163–174.

Etherton, J. and Sneckenberger, J. 1990. A robot safety experiment varying robot speed and contrast with human decision cost. *Applied Ergonomics* 21(3):231–236.

Helander, M. and Karwan, M. 1987. Human motor reactions to dangerous motions in robot operations. PB87-222196, Springfield, VA: National Technical Information Service.

Leamon, T. 1980. The organization of industrial ergonomics—a human: machine model. *Applied Ergonomics* 11(4): 223–226.

Millard, D. 1989. An in-situ evaluation of a capacitive sensor-based safety system for automated manufacturing environments. Technical Paper MS89-301, Dearborn, MI: SME.

Nagamachi, M. 1988. Ten fatal accidents due to robots in Japan. *In Ergonomics of Hybrid Automated Systems I,* eds. W. Karwowski, H. R. Parsaei, and M. R. Wilhelm. Amsterdam: Elsevier.

NIOSH. 1984. *Request for Assistance in Preventing the Injury of Workers by Robots,* HHS Publication 85–103. Washington, DC: NIOSH.

OSHA. 1989. Control of hazardous energy source (lockout/tagout). OSHA Standard 1910.147. *Federal Register* 54(169).

Office of Technology Assessment. 1984. *Computerized Manufacturing Automation.* OTA-CIT-236, Washington, DC.

Sanderson, L., Collins, J., and McGlothlin, J. 1986. Robot-related fatality involving a

U.S. manufacturing plant employee: case report and recommendations. *Journal of Occupational Accidents* 8:13–23.

Sneckenberger, J., Kittiampon, K., and Collins, J. 1987. Interfacing safety sensors to industrial robotic workstations. *Sensors* April.

Tian, J. and Sneckenberger, J. 1988. Performance evaluation of three pressure mats as robot workstation safety devices. In *Ergonomics of Hybrid Automated Systems I*. eds. W. Karwowski, H. Parsaei, and M. Wilhelm. Amsterdam: Elsevier.

# 3

# Industrial Practices for Robotic Safety

Klaus M. Blache

## 1. INTRODUCTION

The current U.S. robot population installed is estimated at about 40,000 by the Robotic Industries Association in Ann Arbor, MI. Data from the International Federation of Robotics shows a decreasing rate of installations in the United States since 1985. This parallels a growing industry trend of wanting to install the proper level of automation (balancing the sociotechnical and manufacturing needs of maximizing the number of robots).

Robots have benefited industry in productive uses removing employees from hazardous work and transfer of technology to other industrial operations. However, robot applications potentially pose new risks to workers that are unlike those associated with conventional material-handling devices and machine tools. For example, although a robot may appear to be idle, a signal from a remote location can cause sudden motion. Similarly, a robot apparently following a repetitive routine may suddenly change that routine in response to a change in conditions on the line. For instance, if one robot on a line is taken out of service, the remaining robots may alter their programs and their path of motion to pick up the missing jobs.

These kind of unexpected robot movements particularly endanger employees who are required to be in a robot's operating work envelope when power to the robot is available. The work envelope is the maximum design reach including the robot end effector (end-of-arm tooling) and workpiece. That portion of the work envelope that the robot actually uses while performing its programmed motions is the operating work envelope. The employees that work in the robot's operating work envelope include personnel who program the robot's motion by manually moving the robot arm from the point to point and maintenance employees who troubleshoot or make adjustments within the work envelope while power to the robot is available. Some employees may actually work beside the robots, handing off or removing workpieces. (Lauck, 1986)

## 2. SAFEGUARDING

It is a recognized fact in the safety engineering profession that most of the industrial accidents are the result of unsafe acts; for example, an improperly trained operator or programmer inadvertently activating the wrong controls or programming motions in an improper sequence. Other accidents result from unsafe conditions (computer or sensor failure, causing the robot to move unexpectedly, or improperly, or mechanical failure on a component member). The types of unsafe acts or unsafe conditions are similar to those of other machines. However, unlike other machines, robots are not captively designed for a specific task with all necessary motion preengineered into their structure. Their design centers on motion flexibility, and therein lies the major area of concern in the safety of all personnel involved with industrial robots (Bretschi and Dirks, 1980).

1. An industrial robot can have simultaneous motion in up to *n* axes, whereas a conventional machine will generally have simultaneous motion in one or two axes.
2. Robots can be programmed for different speeds on each individual axis. Conventional machines have preset, prescribed speeds.
3. Motion of each axis can be freely programmed for a robot, whereas the motion for a conventional machine is prescribed and constrained by mechanical design.
4. Robots generally have a very large motion space compared to the device volume, as opposed to conventional machines, where the machine volume is generally larger than the motion space.
5. The sphere of operation of a robot intersects the working space of other machines and structures, whereas with conventional machines this rarely happens.

Because of these unpredictable motions, it is not normally easy for an observer to foresee the motions of the robot structure. If there is a component failure or a programming error, the motions can be totally unexpected and possibly dangerous because robots have a wide kinematic range of activity. Another safety problem involves the interlinking of other machinery with the robot, where potential failures of this coupled equipment decrease the overall reliability of the entire system. The net result could be erroneous control signals to the robot, causing it to stop or go at the wrong time or to proceed to the wrong position (Bretschi and Dirks, 1980).

## 2–1. Safeguarding for Robots

Having assessed the potential hazards associated with robots, it is now necessary to examine the possibilities to better protect the personnel and equipment. It is difficult to go to a risk-free environment because there is always the possibility of a malfunction, the deliberate acts to defeat safety devices, or the violation of good safety practice. However, there are ways to minimize the potential for these errors.

## 2–2. Training

Of all the assets an organization has, its people are most important. The ability of the human to act or react to an infinite variety of situations makes him or her superior to any other species in existence. However, he or she must have the proper information to ensure that he or she reacts in the correct manner. This is why training is essential whenever a new device or system is incorporated. If he or she acts out of ignorance or fear, the improper response can often occur.

The question of who, how much, and when to train must be an important consideration in the decision to incorporate industrial robots (Trouteaud, 1979).

Obviously, it is necessary to train the programmers or engineers responsible. These people are the prime candidates for training. However, it is also important to make sure that set-up and maintenance personnel and the operators have a working knowledge about the operation of industrial robots.

Programmers must have a thorough understanding of all robot motions and what the controls and program sequences do. While programming, it is important that they watch at all times what the robot is doing. Wherever possible, the programmer must make sure that he or she and other personnel are outside of the sphere of operation of the robot. Even when the robot is turned off, the programmer must be aware that the arm can drift downward, especially on pneumatic or hydraulic systems, as the pressure decays. The manner in which the programmer or engineer performs around the robot is essential. By setting a good example in utilizing safe work practices, he or she will convey that message to others working around the system.

Maintenance personnel must be instructed in effective troubleshooting methods for maintenance and repair. Their ability to find the source of the problem quickly will reduce the hazard potential by limiting the trial-and-error process that can go on blindly if no guidance is given. Finding the source of the problem quickly also reduces the risk of malfunction to others who may be exposed (Trouteaud, 1979).

Wherever possible, all work must be performed with the robot actuators deenergized and the sources of energy locked out. Other maintenance safety procedures must also be followed concerning climbing, lifting, and electrical work.

Finally, operators must have some degree of familiarity with the robot. They must understand where the robot can reach and what it is supposed to do. If part of their job content includes starting of the equipment, they must know the proper sequence for start up and shut down. They must know who to contact in the event of a problem or malfunction. Familiarity with the devices used to stop the robot is essential. Above all, operators must be instructed to stay out of reach of the robot at all times and what safeguards are provided for their safety.

## 2–3. Safeguarding in Mechanical Design

Proper safeguarding depends very heavily on the reliability, strength, and structural characteristics of the equipment. Industrial robots must be designed with the necessary mechanical strength to withstand the normal everyday use and abuse to which it is exposed. Proper selection of materials and heat treat, along with appropriately applied safety factors, are necessary to ensure the reliable operation of the equipment in its planned environment, taking into account the load and task characteristics, the chemical environment, and temperature extremes that the robot will be working in. It is recommended that manufacturers specify the maximum recommended working load that the robot system can handle, along with a listing of environments for which the robots are or are not recommended (Bretschi and Dirks, 1980).

Sharp edges and corners, especially on the outer axes extremes, are to be avoided to prevent injury in the event of contact with a robot component. Control lines—electrical, pneumatic, or hydraulic—should be protected wherever possible to avoid exposure to abrasion or severance as the robot functions. This will help to minimize potential for loss control.

Design for reliability is essential for safety. Consideration should be made regarding the type of control lines used to ensure that the necessary amount of flexibility, abrasion resistance, and power load capacity is provided.

Design for reliability is essential for safety and productivity. By reducing the downtime through reliability, a higher degree of safety can be achieved, especially for the maintenance personnel who along with the programmers have the highest degree of exposure to a hazardous situation.

Wherever there is exposure to mechanical motion (drives, gears, belts, couplings, pulleys, and linkages), guards should be provided (Bretschi and Dirks, 1980).

## 2–4. Controls

Controls must have a high degree of reliability to ensure their continued proper performance in operation. System redundancy is a powerful tool in increasing the reliability of a device. However, complete redundancy can be expensive and cumbersome in the overall system operation. Kuka, for example, has a completely redundant safety function in the robot, where a separate section in the microprocessor monitors the program and motion of the robot. In the event of a failure in either the main motion control system or the safety monitoring system, the robot will shut down immediately (Worn, 1980).

The safety section acts directly on the driving power to the axes, where if an emergency stop button or other safety device is activated, the driving power is shut off independent of the control logic. This safety section should process these conditions (Bretschi and Dirks, 1980):

1. Emergency stop buttons.
2. Safety interlock devices.
3. Cancellation of protective equipment for set-up with the requirements of a deliberate action to select.
4. Selection of the operating modes (off, program/set-up automatic).
5. Axis speed and distance limitations.
6. Slow motion for set-up.
7. Prevention of start up in automatic mode with safety equipment canceled.

Control layouts must be clearly labeled and arranged in such a way to minimize inadvertent actuation of the wrong axis. Depressed motion control buttons with constant pressure would be desirable on teach-pendants or other programming devices to prevent activation through bumping and to allow immediate stoppage of motion if the buttons were released. Emergency stop buttons must be on all stations where motion of the robot is controlled. If more than one programming station is present, a control station selector switch should be used to isolate control of all robot motion to one station. However, emergency stop buttons must always be operational, regardless of the station selected.

If the power supply is interrupted, the control function must guarantee that automatic restart cannot take place if the power is resumed. All safety interlocks must stop the robot as quickly as possible if activated. If either the power supply is interrupted or an interlock device is activated, restart must only take place through deliberate activation of a control selection.

For start up of an industrial robot, it may be desirable to utilize a warning device, visible or audible, prior to the initiation of robot motion. The pause before start up would warn personnel in the area of the pending movement of the robot.

Braking devices are necessary on some robots to ensure that the robot can stop its motion either in its normal operation or if an emergency stop device is activated. A number of different types of braking devices are available:

1. Electromechanical.
2. Reverse current braking.
3. DC braking of AC motors.
4. Hydraulic.
5. Resistance.

In the event that motion stoppage is required, all of these systems will stop the motion. The electromechanical brake will stop the motion and keep the brakes applied by means of a mechanical force such as a spring holding the shoe against the friction plate, disk, or drum. Even with the power off, the brake will hold the position of the axis, thus preventing collapse of the robot arm. Reverse current braking slows the motion by reversing the current to the drive motor, thus providing an opposing torque to the inertial torque. However, there is no protection for collapse of the arm. Hydraulic braking takes place through the restriction of reversal of fluid flow to or from the cylinder or motor. Here again there is no protection against collapse of the robot. Resistive braking works by turning the activating motor into a generator and applying a load across the line to dissipate the energy being generated. Here also there is no protection against arm collapse. The best braking system to use is the electromechanical system because, in the event of power loss, the brake will activate and hold the position of the robot (Bretschi and Dirks, 1980).

Another means of preventing the arm from dropping into a hazardous position due to loss of power is through the use of a blocking or shot pin. In the event of power loss, the shot pin will move into place by means of spring action and prevent the arm from falling any further once it has come to rest on the stop. The effectiveness of this device, however, is limited because it will only work if the arm is working at a height above the shot pin (Bretschi and Dirks, 1980).

Since robots in most cases do not operate throughout their entire sphere of operation, some additional safeguards can be applied to prevent movement beyond preset extremes. Mechanical stops can be preset, which would prevent the robot from moving any further. However, this abrupt stopping could result in damage to the drive system or breakage of the mechanical stop. Limit switches on the axes could also be preset so that in the event of overtravel, the robot would stop. A final method would be through the program controller itself where if the feedback devices (tachometer and resolver or encoder) sense a violation of preprogrammed position time and speed limits, the system shuts down. A combination of at least two of the three systems is the most desirable because of the redundancy.

Whenever a robot is mobile, it is necessary to provide a device which would deactivate the robot in the event that it strikes an object. It is highly desirable to presense the pending collision to allow for stopping time. A number of devices are available to do this, for example, radio-frequency fields, ultrasonic devices, and photocells, all with varying degrees of reliability and sensing capability. Nevertheless, an impact-absorbing material would be beneficial to minimize injury or damage due to the striking of a person or obstacle. The features can also be used on stationary robots to some extent to sense the presence of a foreign element in the area and minimize the effects of a collision (Houskamp, 1981).

To ensure that the robot can only be started in the true home position, an independent device such as a reflector and photocell/light source combination can be used. By attaching the reflector to the arm of the robot and locating the light source/photocell near the true home position, a second, positive check can be made. If the robot is not at that position, then it cannot be started up in the automatic mode.

It may also be desirable to have an automatic slow speed control for programming and maintenance work. By reducing the power to 10%, the hazard is greatly reduced in the event that personnel must work within the sphere of operation while the robot is moving. However, this feature may not be possible if it is necessary to pace the robot motion with other operations. The programmer can also lose the concept of the overall motion if the speed is reduced. By and large, it is highly recommended that a powered down situation be used if personnel are required to be in the sphere of operation (Worn, 1980).

Whenever the operation of a robot must be synchronized with the operation of other equipment, it may be desirable to have both devices automatically start up simultaneously. This can be done provided that personal safety is not jeopardized in any form. All protective and stop devices must be operational before this can happen. A means can also be provided so that the equipment can be run individually, if necessary. Any activation of the safety devices must result in the shut down of all equipment in the safest possible manner. Because of this, advance planning must be done with each installation.

Whenever an operation can be done with either a human or a robot, interlocking features must be provided to ensure that the robot cannot be activated if the manual operator mode is selected. In addition, necessary safeguards, such as two-hand controls, light screens, or safety gates that protect the operator from the hazards of the point of operation, should automatically become functional when the manual operator mode is selected.

A final point to consider for control is the gripper or manipulator itself. The gripper must be designed to ensure that it can provide the necessary holding force under static and dynamic conditions. It should be considered also if it is more beneficial from a safety standpoint for the gripper to release the object rather than hold the object if the power is interrupted.

## 2–5. Safeguarding in Layout

Proper layout is necessary to provide an adequate work area around the robot. The first consideration should be the spatial requirements for the operation of the robot. Consideration must be made to provide the necessary clearance for safe and efficient programming, maintenance, and operation. Surrounding structures, interlinked equipment, and operator stations must be considered to ensure that the possibility of personnel being struck or pinned by the robot or other elements is minimized.

One way to guard a robot is to fence it in. This will keep all personnel from entering the sphere of operation. In the event that it is necessary to enter, interlock devices such as safety plugs or limit switches on doors or gates, automatically shut down the system, or freeze motion. Fencing is called direct guarding because it is not possible to reach over, under, around, or through the device into the hazardous area. Care must be taken, though, to ensure that the guard does not create a hazard itself by creating a pinch point. Fencing protects in two ways. First, it prevents access to the point of hazard, and second it protects personnel from flying objects in the event that the robot should malfunction. Fencing should not generally be considered as a means to contain the robot motion because the force with which a robot can move is much greater than what the fencing can withstand (Bretschi and Dirks, 1980; Trouteaud, 1979).

Indirect methods of guarding can also be used. Guarding of this type does not prevent entry into the area, but should personnel cross the line of demarcation, the system would be deactivated automatically. Examples of this system would be light screen, radio-frequency curtains, pressure-sensing floor mats, and profile gates (devices in conjunction with limit switches that when moved stop the equipment).

There is the possibility of accidentally stopping the robot by walking through the invisible barrier unintentionally. While this is still safe, productivity can be hampered. It would be desirable to at least paint lines on the floor, marking out the safeguarding areas. Caution should be taken with these devices to make sure they are adjusted properly. They should be designed and installed so a single component failure or ambient condition, such as flashing lights, EMI (electromagnetic interference), or RFI (radio-frequency interference), will not reduce their safeguarding effectiveness.

With multiple robot set-ups, each unit should have its own safeguarding features because it may be possible to shut down each robot individually for programming or maintenance. Safeguarding devices and work practices should be established so that in no case will a person be injured by the operation of an adjacent robot or machine whenever servicing is required.

Wherever operators work adjacent to a robot station, consideration must be given to ensure that the workstation does not overlap into the sphere of operation.

If this is necessary, safeguarding devices, such as light screens, safety mats, movable barriers, or cycle initiation button can be used to ensure that there is no hazard when the robot is in the vicinity. Direct contact with the robot gripper should always be avoided unless tools or other safeguards are provided to ensure that the manipulator will not grasp the operator and pull him or her into a dangerous position. Conveyor belts, magazine or track feeds, or drop chutes can be used as alternatives to direct parts transfer and loading (Bretschi and Dirks, 1980).

In planning the layout, it is desirable to keep the normal operating sphere as small as practical. This aids in safety because if the distance is smaller, less inertia is generated by the robot. There is also a reduction in wear and tear on the robot. The amount of guarded area is reduced, resulting in a cost savings. Smaller moves also result in improving cycle times.

In the layout and design of fixtures, benches, and interlinking equipment, minimize the number of obstructions such as hoses, lines, and protrusions that could be struck by the robot. Make sure also that, wherever possible, these devices are securely fastened to prevent their movement, if they should be struck by the robot. Proper layout should also provide adequate visibility of the work area so that observation of the ongoing processes can be done without entering the sphere of operation. Wherever possible, process equipment should be laid out so that there is a clear, observable separation of processes, thus aiding the programmer in his or her work. If visibility is difficult to attain, mirrors or closed-circuit television might offer a solution (Bretschi and Dirks, 1980).

## 2–6. Safeguarding in Programming

In some cases, it is necessary temporarily to cancel safeguarding devices to program the robot. However, there are measures that can be taken to reduce the possibility of an accident:

1. Designate a safe position for programming.
2. Reduce power during programming. 250 mm/sec is the accepted speed (ANSI and ISO).
3. Provide clear visibility of the robot function during programming.
4. Provide emergency stop control on the teach-pendant.
5. Use constant pressure control to cause robot movement.
6. Avoid pinch points within the sphere of operation where the programmer is likely to be.

More specifics on safeguarding when in a teach mode can be found in ANSI RIA R15.06, Section 6.4.4. Some of these items may or may not be feasible depending on the application. The prime safety factor though is the total control and awareness of the programmer with regard to the robot movement. These other measures enhance the safety factor but could be for naught if the programmer is not paying attention to the task at hand, or if he or she should lose control of the robot.

## 2–7. Safeguarding in Maintenance

As with programming, it may be necessary to be in the sphere of operation while the robot is moving for maintenance purposes. Many of the safeguarding procedures that apply to programming apply also to maintenance. A number of additional items can be provided to enhance the degree of safety in maintaining and repairing robots.

If power is not necessary to do the work, then an effective, established lockout procedure must be used. This will ensure that the equipment cannot be accidentally started up.

Stands to prop up the arm can be used to prevent the arm from falling while work is being performed. Motion can be isolated to only certain axes by isolating the energy only to those needed for maintenance. Shackles can also be used to restrain the motion of the robot so that in the event of inadvertent motion, maintenance personnel are not endangered.

Diagnostic displays such as indicator lights, computer printouts, and CRT displays can be a major asset in maintenance safety. By using displays outside of the hazardous area, malfunctions can be diagnosed safely and quickly, minimizing exposure time and downtime (Worn, 1980).

Preventive maintenance with adequate stocking of spare components will first of all prevent major failures of components, thus providing an additional safety measure. Second, by minimizing major failure, the overall hazardous exposure to maintenance personnel is reduced. In some cases, it may be beneficial to have a robot removed from service for necessary repair. By doing this, work can be done in a less confined area, outside of the hazards associated with the robot's working environment (Bretschi and Dirks, 1980).

General safety rules for skilled trades personnel must always be followed. Safe practices for lifting, proper usage of tools and equipment, and repair procedures must always be followed. All safety devices must be put back into operation once the job is complete. If a guard is designed with removal and replacement in mind, chances are it will be put back on once the job is complete (Cincinnati Milacron, 1981).

## 2–8. Installation

Once a robot is installed, it is a good idea for the responsible safety personnel to look at the installation, to check for any other unforeseen hazards, and to ensure that all safety devices are in place and functional. There are few jobs that can be fully thought out so that all problems are anticipated before installation. If there are serious problems, correction must take place before the equipment is released for production. The information discovered in this review can be invaluable in making sure that the same mistake is not made twice, resulting in additional cost and lost production time (Bretschi and Dirks, 1980).

Safety personnel should be involved before, during, and after the installation. Their advice can reduce the risk of an accident with the robot. By performing periodic safety checks, the continued reliability of the safeguarding devices can be ensured.

## 2–9. Where Do We Go from Here?

Continued development in robotics can provide an additional degree of safety. Development of preception for robots in the form of obstacle sensing can only improve the safety margin. If a robot can sense an obstacle before it strikes it or if it can be stopped before serious injury or damage takes place, the overall safety environment for robots is improved. Development of senses such as sight, touch, and hearing can be an asset to the safe operation, for if the equipment itself can recognize a potential hazard and then react in a safe manner, the risk for injury is greatly reduced.

## 2–10. Conclusions

Robots are an inevitable part of the future of industry. Their increased utilization will result in increased productivity and a safer work environment for all concerned. However, their successful implementation will only happen if their use will only benefit and not harm the human.

Socially, there is great potential for changing our way of life. However, it will only happen if the change improves the quality of life. If mass unemployment results, then we have gained nothing and lost much, and the new technology will cease to exist. However, if the standard of living improves, the new technology will flourish and grow.

A major portion of this chapter was dedicated to the safety aspects of robots. Safety must always be a prime concern, for if accidents happen with this equipment, fear and rejection will result. Safeguarding the operation of robots in the form of proper procedures, layout, design, and controls are essential in the application of this growing technology.

Robots designed today are generally safe devices because of our growth in the understanding of safety and equipment. By systematically analyzing the hazards and assessing the risks, viable solutions can be found that reduce or eliminate the hazards in a manner that is still cost-effective. Concern for the company's most valuable resource—its people—must always be a high priority, for that continuing concern will keep our industry alive and healthy (Herlache, 1981).

## 3. INDUSTRIAL EXAMPLES

The following section presents information from Packard Electric, Inland Division, Delco Electronics, and Delco Moraine regarding guidelines and examples involving robotics safety.

### 3–1. Packard Electric

Packard Electric robot cell safety systems incorporate redundancy. A number of the robot work areas include a perimeter mesh fence with entry through electrically interlocked gates (Figure 3–1). The redundancy feature is usually accomplished through pressure-sensitive floor mats within the envelope. If either is violated, the power source for the motion is removed. From an operational perspective, it is wise not to drop controller power. Maintaining controller power permits the joint encoders to hold their positional values; this eliminates the need to recalibrate. Maintaining controller power also removes the need to reload the

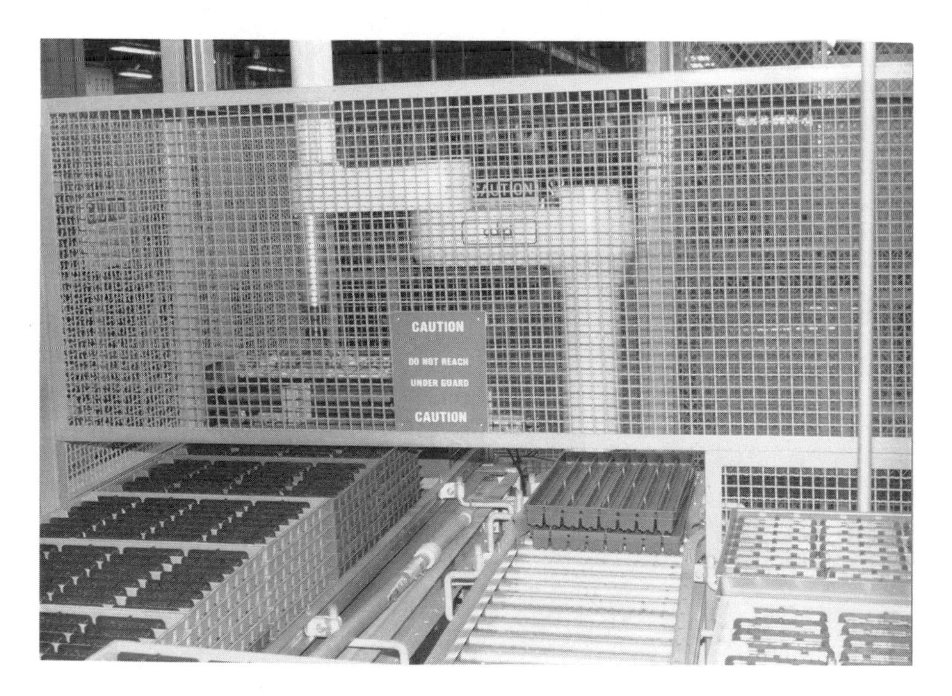


**FIGURE 3–1.** Mesh safety gate.

program into memory. One of their cells includes a pair of light curtains as the perimeter guard for a table-mounted robot. The light curtains combined with pressure mats initially resulted in frequent nuisance interruptions to the workcell, as visitors would inadvertently either step on a mat or extend their hand through the light curtain (Figure 3–2).

Next are safety guidelines for robot installations. They should be used in the planning of each robot operation by the engineers responsible for the project. Although each line must be unique to a certain degree, any deviation from the principles herein should be brought to the attention of the Safety Department for approval at an early stage of design. The purpose is to establish guidelines for the safe construction, care, and use of industrial robot systems. These guidelines are intended to protect personnel and provide them with safe access to the robot system.

The actual guidelines begin with definitions, most of which were adapted from the Robot Institute of America (RIA). The definitions are provided at the end of this section.

**FIGURE 3–2.** Light curtains for safety.

***Installation***

General construction and equipment installation will be made so as to restrict all employees from entering the operating envelope except when teaching, programming, servicing, and/or performing maintenance. The above are the only authorized entries. Safeguards with at least one level of redundance will be utilized to ensure no unauthorized entry into the work envelope. Software limits will not be considered as a safeguard. The following principles should be adhered to in making a safe installation:

**1.** The project engineer will work with the robot manufacturer and Packard Safety to eliminate by design as many human hazards as possible. Safety must be contacted to review design drawings and safeguard plans to release the robot for debug and to release for production.

**2.** Any modifications to existing processes involving robots will be reviewed by engineering and safety. When necessary, the manufacturer will be consulted.

**3.** There shall be adequate clearance between the robot and any building structure. This will allow for human clearance between the restricted work envelope and any building structure.

**4.** The primary electrical control panel shall be located outside of the work envelope where the robot can easily be seen. This panel and all operator control panels will be equipped with emergency stop buttons. Deliberate action on the part of the operator will be necessary to restart the robot after an emergency stop button has been activated.

**5.** A lockout/tagout capability shall be provided.

**6.** Stored energy sources shall be labeled and a means for releasing it safely will be provided.

**7.** Electromagnetic and radio-frequency interference will be minimized by design.

**8.** Actuating controls shall be designed and labeled to prevent inadvertent operation. Key selector switches, teach cycle mode switches, and two-hand controls are other possible means of accomplishing this.

**9.** The teach-pendant for use within the work envelope will be equipped with an emergency stop button. Release of pendant controls shall stop all motion, and it will be impossible to place the robot in automatic operation using the pendant. Unless specifically required, the maximum slow speed at any axis will be 255 mm/sec.

**10.** Variations in electrical, hydraulic, or pneumatic pressures shall not create any hazardous motion of the robot.

**11.** Environmental conditions affecting employees and the robot will be evaluated upon a worst case anticipated level.

**12.** Amber lights visible from any point of entry into the work envelope will be provided. The lights will be on when there is power to the robot and will flash when there is power to the actuators.

**13.** The use of hydraulics is discouraged. If used, they will be of the nonflammable type.

**14.** When barriers are not used to enclose the entire work envelope, warning signs will be placed at locations where entry could be made. Also, other electrical safeguards will be installed at the point of entry to stop the robot and place it in a hold position.

**15.** Gates used in conjunction with barrier guards will be electrically interlocked. After closing an opened gate, the main control panel must be activated before the robot can be put in automatic mode.

**16.** Fencing will be constructed at such a height so as to discourage employees from climbing over or crawling under the fence. It shall be painted flat black. Clear acrylic sheet may be used for increased visibility.

**17.** Yellow paint under removable barrier guards will be utilized to indicate any barrier that has been removed.

**18.** Mechanical stops will be able to withstand the force of momentum of the robot traveling at maximum speed and carrying a full load.

**19.** When drive power is interrupted to the robot actuator, the tool grip will be maintained. Any other engineering steps to prevent materials from being dropped or thrown will be taken prior to installation.

**20.** If an unexpected interruption to the teach or normal work cycle occurs, power to the robot actuators shall be interrupted. A manual reactivation of robot actuator drive power, from outside the envelope, will be required after identifying the problem, safely eliminating the cause of the interruption, and repositioning the robot to the place where the normal work or teach cycle can again begin.

**21.** Adequate engineering controls such as light curtains, pressure-sensitive mats, or a shuttle mechanism with two-hand controls will be taken to prevent the possibility of any employee–robot crossovers when the robot is in the normal cycle mode.

**22.** When barrier guards are not utilized, guard rails, pressure-sensitive mats, floor color coding, and other safeguards should be so designed to prevent employees from reaching or entering into the work envelope during normal cycle.

***Training***

**1.** All engineers, technicians, tradesmen, operators, or other employees working in a side-by-side relationship to the robot will receive job safety instructions from their supervisor before beginning work.

**2.** Vendor safety training programs will be utilized where available and applicable to the Packard installation. Manufacturer's safety rules will also be reviewed with appropriate employees.

**3.** Individualized training for each robot installation should be given concerning testing/start-up procedures, teaching the robot, playback, computer utilization, maintenance, and service. It is supervision's responsibility to ensure safety-related information concerning any of the above is given to employees prior to beginning work.

**4.** All employees will be trained to report any safety-related incidents involving robot operations. Supervision should complete a "robot and robot system incident report" and forward it to the Safety Department within 3 days of the incident. The report covers the areas of robot identification, type of application, incident description, actions taken to prevent recurrence, and comments. Any injury involving a robot should be reported immediately.

The following rules are for safe use of robots, production and teaching or maintenance considerations.

***Production Considerations***

**1.** If the robot is not moving, DO NOT assume it is not going to move.

**2.** If the robot is repeating a pattern, DO NOT assume it will continue.

**3.** Always be aware of where you are in relationship to the possible positions that the robot may reach.

**4.** Be aware if there is power to the actuators. Indicator lights will be on when there is power to the actuators.

**5.** Limit switches or software programming will not be used as the primary safeguard.

**6.** Teaching, programming, servicing, and maintenance are the only authorized reasons for entry into the work envelope.

**7.** Safeguards with at least one level of redundancy will be present when employees are required to enter the work envelope: that is, power is off and pressure-sensitive floor mats are covering the work envelope, or a gate with limit switch is open and pressure-sensitive floor mats are covering the work envelope.

**8.** Never climb on, over, or under a barrier for any reason.

**9.** Before activating power to the robot, employees should be aware of what it is programmed to do, that all safeguards are in place, and that no foreign materials are present within the work envelope.

**10.** Eliminate any source of stored energy prior to entry into the work envelope.

**11.** Notify supervision immediately when any unexpected interruption to the normal robot work cycle occurs.

**12.** Servicing should never occur within the work envelope with power on to the robot.

**13.** Report any missing or defective safeguard to supervision immediately. Check all safeguards at the beginning of each shift.

### *Teaching or Maintenance Considerations*

**1.** Teaching, whenever possible, should be accomplished outside the work envelope.

**2.** Whenever possible, only the employee utilizing the teaching pendant should be inside the work envelope. Another employee should be present outside the work envelope with full visibility of the teacher and the robot arm. The main control panel should provide the best view.

**3.** While operating the robot with the pendant, eye contact must be maintained while the robot is moving and both hands should be in contact with the pendant.

**4.** Check that the emergency stop on the pendant is functional before entering the work envelope.

**5.** A pathway under or away from the work envelope should be present whenever possible before teaching or power-on maintenance work is performed. Avoid placing any part of your body between the robot arm and any immovable object.

**6.** With power-on and two employees inside the working envelope, the pendant must be used and held by the employee closest to the robot arm. The other employee should be behind the one holding the pendant. Whenever possible, not more than one employee should be within the work envelope.

**7.** Tradesmen must be thoroughly trained in the proper use of the teach-pendant before they are required to work inside the envelope with power-on to the robot actuators.

**8.** The speed of the robot arm and end effector should be relative to the clearance and position of the employee utilizing the teaching pendant. Slower speeds allow for safer operation.

**9.** Whenever possible, replay will occur with gates closed and employees standing outside of the work envelope.

**10.** When no power is required, tradesmen should lock out the main electrical panel before performing maintenance inside the working envelope.

**11.** A second employee should be located at the main control panel anytime that power-on maintenance is being performed inside the work envelope without use of the teach-pendant.

### *Definitions*

**Automatic Operation:** That time when the robot is performing its programmed tasks through continuous program execution.

**Barrier:** A physical means of separating persons from the robot.

**Control Software:** The inherent set of control instructions that defines the capabilities, actions, and responses of a robot system. This software is fixed and usually not modifiable by the user.

**Drive Power:** The energy source(s) for the robot actuators that produce motion.

**Emergency Stop:** A method using hardware-based components that overrides all other robot controls and removes drive power from the robot actuators and brings all moving parts to a stop.

**End Effector:** Part of the robot that holds the tooling at the end of the arm.

**Hold:** A time period when the robot hardware is motionless, but there is power to the robot and the actuators. The robot may automatically move from the hold.

**Industrial Robot:** A reprogrammable multifunctional manipulator designed to move material, parts, tools, or specialized devices, through variable programmed motions for the performance of a variety of tasks.

**Industrial Robot System:** An industrial robot system includes the robot(s) (hardware and software) consisting of the manipulator(s), power supply, and controller; the end effector(s); any equipment, devices, and sensors the robot is directly interfacing with; any equipment, devices, and sensors required for the robot to perform its task; and any communications interface that is operating and monitoring the robot, equipment, and sensors.

**Interruption:** A time period when the robot hardware is motionless because the robot system fails or is disabled. There may or may not be power. Usually an employee must act to eliminate the interruption.

**Limited Devices:** A means for restricting the work envelope that will stop all motion of the robot independent of the control software or the application programs.

**Operating Work Envelope:** The volume of space enclosing the movement of robot manipulator, end effector, workpiece, and the robot itself when performing programmed motions.

**Pendant:** A portable device by which a person can control robot motion.

**Presence-Sensing Device:** A device designed, constructed, and installed to create a sensing field or area around a robot(s) that will detect an intrusion into such field or area by a person, robot, etc. Some examples of presence-sensing devices are light curtains, mats, capacitance systems, proximity detectors, and vision safety systems.

**Programming:** Providing the instruction(s) required for the robot to perform its intended task.

**Proximity Detector:** A device which senses that an object is only a short distance away, and/or measures how far away it is from the robot.

**Repair:** To restore the robot system to operating condition after damage or malfunction. Work is usually performed by skilled trades and/or engineering.

**Restricted Work Envelope:** The volume of space enclosing the movement of robot manipulator, end effector, workpiece, and the robot itself when the robot is restricted by limiting devices which establish limits that will not be exceeded in the event of any foreseeable failure of the robot or its controls. The maximum distance the machine can travel after the limit device is actuated will be considered the basis for defining the restricted robot operating envelope.

**Robot Actuator:** The cylinder, motor, etc., that directly or indirectly drives the unit; commonly thought of as the arm of the robot minus the end effector.

**Safeguard:** A guard, device, or procedure designed to protect persons from danger.

**Safeguard Redundancy Level:** Combining or arranging of safeguards in an order to provide backup safety features for the robot.

**Service:** To make fit for use, adjust, repair, maintain. Work is usually performed by the operator or service person.

**Software Limits:** A limit imposed by control software or some motion that inhibits travel beyond a point. This point or limit over a working range can be changed by trained personnel.

**Teach:** To move a robot to, or through, a series of points to be stored for the robot to perform its intended task.

**Vision Safety System:** A device, such as a camera, that is designed, constructed, and installed to detect intrusion by a person(s) into the robot-restricted work envelope.

**Warning Device:** An audible or visual device used to alert persons to a potential safety hazard.

**Work Envelope:** The volume of space enclosing the maximum designed reach of the robot manipulator, end effector, workpiece, and the robot itself.

## 3–2. Inland Division

The three general categories of robotic applications/installations are as follows:

1. Manual interface with robot envelope—single entry point.
2. Manual interface with robot envelope—multiple entry points.
3. No manual interface—automatic operation.

All three categories employ the use of "open" guarding concept whenever possible. The greatest safety hazard in a robotic installation is not the prospect of being hit by the robot but of being crushed by the robot, being caught between the robot and an immovable object, such as your safety guarding. The preferred method is to provide a waist-high fence or railing around the robot envelope. This concept also facilitates access for maintenance personnel. The second common characteristic is no machine will resume automatic operation after a person has violated the robot's work zone. To resume operation, a manual restart is required.

***Manual Interface: Single Entry Point***

This situation has the operator separated from the robot work envelope by a light curtain. The output or safety contacts of the light curtain are directly wired to the robot HOLD circuit. When the light curtain is blocked, the robot is on HOLD. When the light curtain is cleared, the robot is ready to be manually restarted.

***Manual Interface: Multiple Entry Points***

This situation is similar to the first in that the operator is separated from the robot work envelope by a light curtain. In this case, there are multiple points of entry, therefore, multiple light curtains. The goal is to allow the robot to work in one "zone" without stopping the robot. This is accomplished by external sensors that monitor the robot's position and selectively enable or disable the associated light curtain.

***No Manual Interface: Automatic Operation***

This situation (Figure 3–3) does not require an operator to interface with the robot work envelope on a per cycle basis. Operator assistance is needed to correct exceptional or unusual circumstances. Access to the robot work envelope is through a gate with a photoelectric eye. Limit switches, joy plugs, and proximity switches are not as desirable; they are easier to cheat. Like the first two examples, entry through the safety gate places the robot in HOLD and requires a manual restart to resume operation.

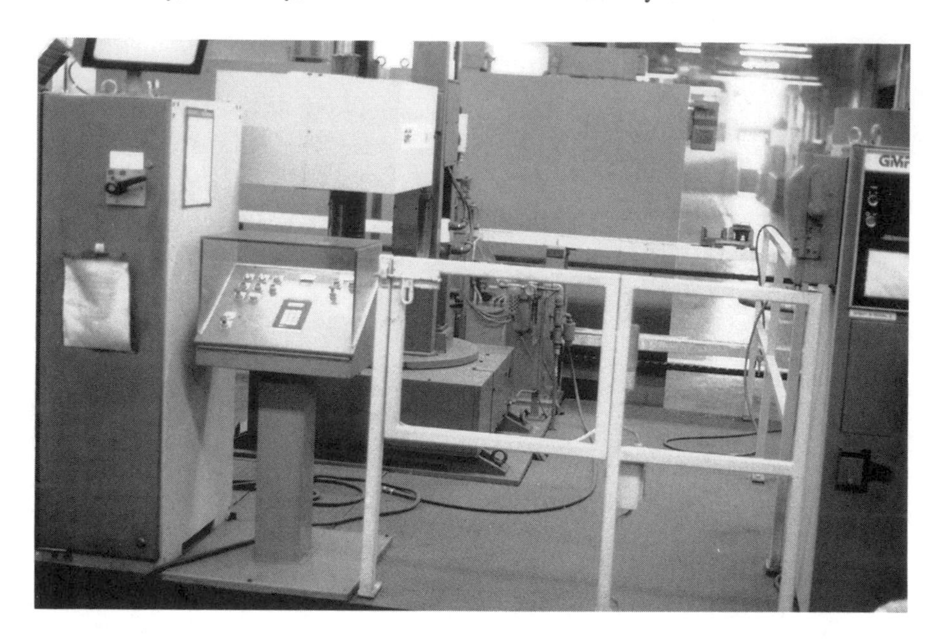

**FIGURE 3–3.** Gate access to automatic operation.

## 3–3. Delco Electronics

The area to be discussed posed a challenging problem for guarding because it is a manufacturing show area. The traditional method of guarding was not acceptable, but reliable guarding was needed owing to the number of people looking at the system and accessibility for the operator. This system is 1 of approximately 10 systems in operation.

The job task is to remove a ceramic substrate from a tray and locate it in a test fixture. It is then tested and reloaded into the same tray if acceptable, otherwise it will be loaded into a reject tray.

A robot much larger than required was utilized for this task owing to the reach that was required. The robot selected was an Automatix's Scara robot. The operation specifics are as follows:

**1.** The substrate tray is transferred by conveyor to the robot unload/load position called the test station. The substrates are mounted vertically in trays with slots to keep them captive. (See Figure 3–4.)

**2.** The robot then takes the substrates out of the tray one at a time and places a substrate on a test fixture (Hewlett Packard test system with dial-type parts shuttle). The test station has two positions, one being tested while the other is

**FIGURE 3–4.** Robot load/unload.

being loaded. If the substrate is a good one, the robot reloads it into the original tray. Otherwise, it is loaded into the correlation parts bin as a reject.

**3.** After the part is tested, either good or bad and placed into its proper place, the robot repeats the load procedure.

Critical criteria for safety system selection:

**1.** The area is a show area of the latest in technology and techniques in parts handling. With that in mind, they did not want unsightly enclosures or obstacles that hamper accessibility to the work area.

**2.** Infrared light curtains would be compatible, but with the obstacles involved, they would not be a good choice. Switch mats were chosen for the safety system owing to the ease of accessibility and their aesthetic compatiblility with the area described. (Figure 3–5 shows the switch mat layout.)

## 3–4. Delco Moraine NDH

The following is the section of the Delco Moraine NDH (DM-NDH) machine control specifications that relates to robotic safety. This is a description of some

**FIGURE 3–5.** Switch mats.

of the safety requirements for DM-NDH robotics cells. Of particular interest are the gate interlock switches. These switches are required to be magnetic or pin-type and cannot be defeated as easily as a standard limit switch. Figures 3–6 and 3–7 illustrate the magnetic door switch location and details.

### 10A Robotic Controls and Systems

#### 10A.1 General

**10A.1.1** Any exception to this specification must be approved in writing by the Delco Moraine NDH Electrical Engineer in advance.

**10A.1.2** The following are acceptable robot controllers:

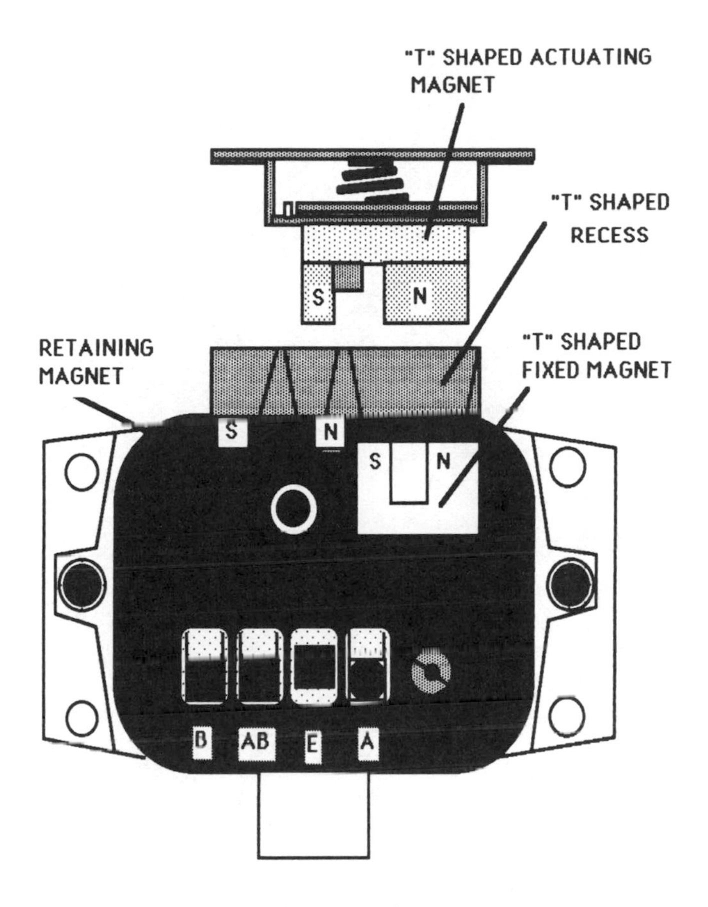


FIGURE 3–6. Front view (cover off) magnetic door switch.

**10A.1.2.1** GMF Robotics R model C with bubble cassette interface.

**10A.1.2.2** GMF Robotics R model F (KAREL).

**10A.1.2.3** All other controllers must be approved, in writing, by Delco Moraine NDH Electrical Engineering.

## 10A.2 Safety

**10A.2.1** Safety of robotic equipment is a high concern for both Delco Moraine NDH and General Motors Corporation. It is important that robotic safety be closely coordinated with the Delco Moraine NDH Safety Department and Electrical Engineering.

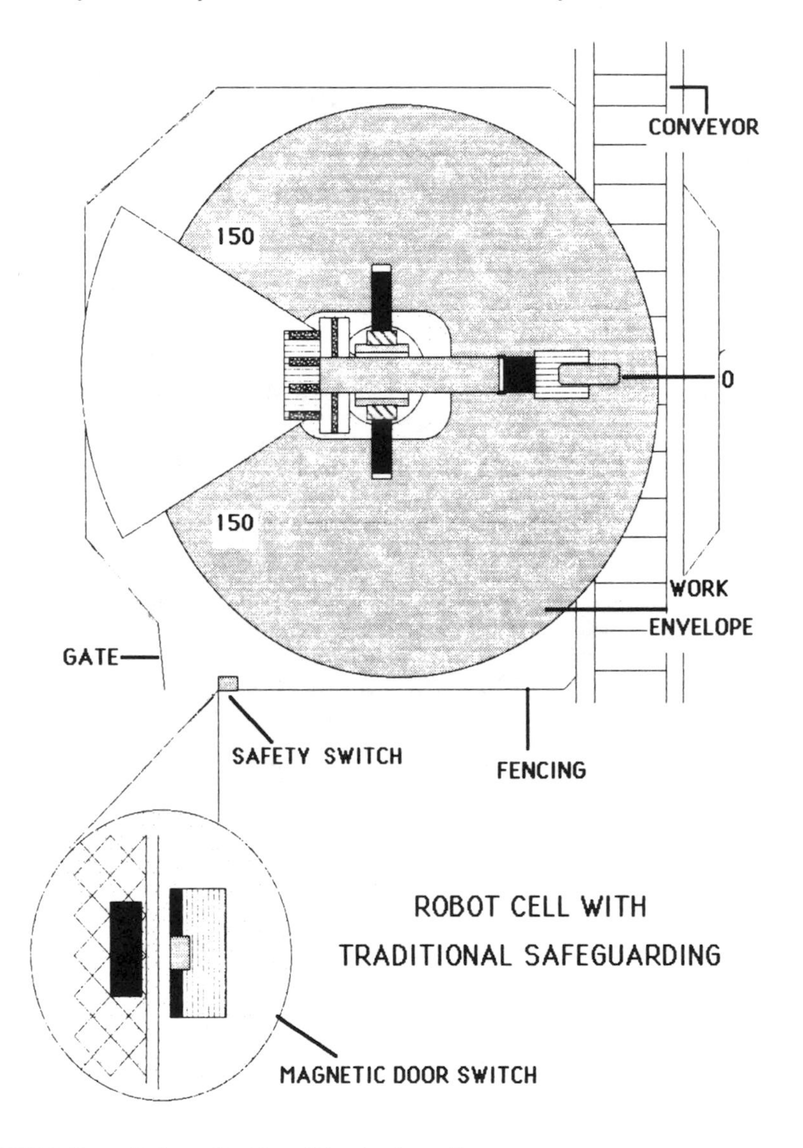


**FIGURE 3–7.** Robot cell with traditional safeguarding.

**10A.2.1.1** If it is determined that the robot must be enclosed in a barrier, penetration of the barrier must place the robot in an Emergency Stop condition and remove power from the robot.

**10A.2.1.2** The following are barriers acceptable to Delco Moraine NDH.

**10A.2.1.2.1** Fencing. The height of the fencing must be approved by Delco Moraine NDH Safety.

**10.2.1.2.2** Light Screens. Shadow [SHAD-IV (NEMA 12)] is approved for this use.

**10A.2.1.3** The following switches are approved for use on robot fence gates.

**10A.2.1.3.1.** Gould Mobrey magnetic switch. Part Number U20800.

**10A.2.1.3.2.** Harrison gate switch. Receptacle part Number 22803 and cap Part Number 22808.

**10A.2.1.4** When the required work envelope of the robot is smaller than the work envelope capability of the robot, mechanical stops shall be used on the base to reduce the work envelope of the robot to the size required.

**10A.2.1.5** The robot controller shall be placed outside of the work envelope of the robot.

**10A.2.1.6** The robot controller shall be connected to the load side of the robot cell power distribution panel. The robot controller shall have a separate disconnect with lockout capability.

## 10A.3 Machine Operator Controls

**10A.3.1** The robot control design shall not require the machine operator to use the pendant control as part of his work. Separate annunciation and recovery control is to be provided. Annunciation, using indicating lights or message displays, shall identify critical conditions of the robot to the operator (i.e., robot fault, robot home, part present, etc.). Robot recovery control from abnormal conditions (other than robot equipment failure) is described in paragraphs 10A.5.4 and 10A.7.2 of this specification. If the robot is not part of a cell where these devices can be included as part of an operator's station, they are to be provided in a separate NEMA 12 enclosure.

**10A.3.2** Delco Moraine NDH Machine Control Specification shall apply to controls and wiring external to the robot and its controller.

## 10A.4 Robot Cell Control (R model C, R model F)

**10A.4.1** The robot controller is to be used to control the robot, the robot tooling, and robot diagnostics. The robot controller is to be used as an I/O interface between the robot and the cell controller. The robot controller is not to be used for the control of other peripheral equipment.

## 10A.5 Robot Software Requirements (R model F)

**10A.5.1** The robot software shall function in a logical manner.

**10A.5.2** Software shall be modular through the use of subroutines. This will make each robot function easy to follow.

**10A.5.3** Labels and variable names shall be chosen so that they are descriptive of their functions (i.e., PICK PLACE as a label for a pick and place routine, PICK as a variable name for the pick up position.)

**10A.5.4** Graceful recovery is to be implemented. Graceful recovery is the process of returning to home position automatically from any programmed point in space through safe recovery zones, using standard operator devices (pushbuttons and selector switches), not the pendant control to initiate this procedure. A safe recovery zone is an area through which the robot can recover from a point or set of points without causing damage to equipment due to a collision. This concept must be approved by the Delco Moraine Electrical Engineer.

### 10A.6 Software Documentation (R model F)

**10A.6.1** The purpose of robot software documentation is to make the robot program easy to follow and understand.

**10A.6.1.1** A program flowchart, using standard symbols is required.

**10A.6.1.2** The main line of the program shall contain a detailed description of the function of the program. It shall also contain a description of all subroutine labels as well as a short description of each subroutine functions.

**10A.6.1.3** A detailed description of the subroutine function is required at the start of each subroutine.

**10A.6.1.4** Line by line descriptions are required when necessary to provide information on how the subroutine task is being accomplished.

**10A.6.1.5** A computer printout of the robot program is required. A backup disk of the program is also required. The disk size shall be specified by a Delco Moraine NDH Electrical Engineer. Robot hardwire documentation, Input/Output information and flowcharts shall be drafted in accordance with Section 5.0 of Delco Moraine Machine Control Specification.

### 10A.7 Robot Software Requirements (R model C)

**10A7.1** Robot Software shall function in a logical manner. To accomplish this the following guidelines are to be followed:

**10A.7.1.1** Robot flag designations shall correspond to N-point numbers (i.e., N010 S97, 10)

**10A.7.1.2** Software modularity, through the use of subprograms accessed by branch instructions from the main program, is required.

**10A.7.2** Graceful recovery is to be implemented. Graceful recovery is the process of returning to home position automatically from any programmed point in space, through safe recovery zones, using standard operator devices (pushbuttons and selector switches), not the pendant control to initiate this procedure. A safe recovery zone is an area through which the robot can move to recover from a point or set of points without causing damage to equipment due to collision. Concept must be approved by the Delco Moraine Electrical Engineer.

**10A.8 Software Documentation (R model C)**

**10A.8.1** Robot R model C software documentation shall contain the following information:

**10A.8.1.1** A program flowchart, using standard symbols, is required.

**10A.8.1.2** The N-point number (N012).

**10A.8.1.3** The G-codes associated with the point (G12).

**10A.8.1.4** The F-Code associated with the point (F6).

**10A.8.1.5** A description of the position associated with the point.

**10A.8.1.6** All S-codes associated with the point (S80, 1, S00).

**10A.8.1.7** The flag number associated with that line (if set).

**10A.8.1.8** A description of all S-code operations.

**10A.8.1.9** A computer printout of the program is required. A backup disk or bubble storage of the program is also required. The disk shall be a 3.5 in. floppy with the capability to run with the GMF documentation package for the R model C controller. Robot hardwire and Input/Output information and flowcharts shall be drafted in accordance with Section 5 of Delco Moraine NDH Machine Controls Specification.

**10A.9 Workmanship**

**10A.9.1** Unless otherwise specified, protection for cables connecting the robot to its controller shall be provided, to prevent damage in the Customer's environment. Support and protection above the floor level is preferred for most installations.

**10A.9.2** Exposed small electrical cables and/or pneumatic tubing lines, used to sense or operate devices on or near the robot, are to be installed in a workmanship-like manner. They shall be neatly dressed and protected from wear points and potential falling parts.

***Discussion***

At DM-NDH a General Motors Fanuc (GMF) robot is used for loading caliper castings on to the assembly line. Truck entrance into the working end of the robot is necessary for the removal of empty dunnage. The open area used for truck passage needs to be protected by a safety system that will terminate the operation of the robot upon human intrusion into the cell. Figure 3–8 depicts the loading operation described, and Figure 3–9 shows the magnetic door switch.

Infrared-sensing technologies are currently being developed for some safety systems to cover the working envelope of the robot. Infrared radiation occurs at

**FIGURE 3–8.** Robotic loading of caliper casting.

wavelengths from 1 $\mu$m to 1mm in the electromagnetic spectrum. The infrared sensor is a passive device that relies upon receiving infrared radiation. The basis for infrared sensors is that all objects emit black-body radiation if their temperature is above absolute zero. It was determined that humans emit thermal radiation that produce wavelengths in the spectral range from 8 to 14 $\mu$m; 10 $\mu$m is peak for a stationary person.

The type of thermal or infrared sensor being used is a lithium tantalate pyroelectric detector. The detector produces an output voltage in response to changes in temperature. Therefore, since humans emit infrared radiation, these sensors can be used to detect human intrusion into the robot's work area (RPI, 1989). To evaluate the ability of the infrared system to detect human movement in a designated area, testing was performed in the training area (in 1987) with a GMF training robot. The robot was programmed to move in a manner that closely simulated the actual movement of the loading robot described above.

It was concluded that the sensors were able to detect humans at a range of 2–10 ft (depending on the sensor). Rensselaer Polytechnic Institute has

**FIGURE 3–9.** Magnetic door switch installed.

been able to detect human intrusion up to approximately 15 ft. However, in the manufacturing environment, noise is still a problem for these sensors.

## 4. CLOSING COMMENTS

In the last decade the drive toward automation has increased rapidly. Currently that movement (or more appropriately, the rate of that trend) is also being questioned because state-of-the-art factories in U.S. industries are not performing as well as planned. I believe that much of this is not directly due to automation technology itself, but rather the proper implementation of that technology with the people impacted. In addition to automation technology, it is also due to issues concerning the following:

- Operators of high-technology equipment.
- Repairmen of high-technology equipment.
- Employees working next to high-technology equipment.
- Supervisors with changed responsibilities due to high-technology equipment.

- Employee training and related learning curves.
- Everyone's work attitude and focus.

The issues revolve around such necessary preautomation tasks as specifying the "proper" level of automation with adequate training, achieving and maintaining goal clarity (common focus), and promoting worker participation to successfully implement the new technology.

In order to regain/maintain/lead worldwide manufacturing competitiveness, companies must become masters at change. It is critical that each employee understand and believe that they, both individually and collectively, can make a difference. It is these persons and teams, taking risks and making continuous incremental improvements that will shape and determine the future position of their companies (Blache, 1988).

The topic of robotic safety is also an issue of successfully implementing change. There are many good safety innovations. However, in sociotechnical terms "change is greater than innovation"—meaning that getting employees to accept and properly use the changes (safety innovations) may be the greater challenge. Better understanding the critical balance between people and technology, the social impacts of automation and technology on people and managing change will have a significant impact on robotics safety during the next decade.

## ACKNOWLEDGMENTS

I would like to acknowledge several persons who contributed to this information: H. Dudley Greeson—Delco Electronics Corporation; Russell L. Herlache—Saginaw Divison; Gregory M. Hickey—Delco Moraine NDH Division; Kenneth E. Lauck—Corporate Occupational Safety; Steven Pyle—Inland Division; and Jerry Shaffer and Jack Gysegem—Packard Electric Division, all from General Motors Corporation. Without their valuable input, this report would not have been possible.

### References

Blache, K. 1988. *Success Factors for Implementing Change*. Dearborn, MI: Society of Manufacturing Engineers.

Bretschi, J. and Dirks, M. 1980. New handling systems as technical aids for working process: Part 7—Sensors and safety equipment. National Technical Information Service. No. PB81-199 242:118–178.

Cincinnati Milacron. 1981. *Owner's Manual for T-3 Robots*.

Herlache, R. L. 1981. The Safety Aspects of Robotics. University of Michigan. I & OE 590 Report. pp. 1–52.

Houskamp, R. W. 1981. Obstacle protection with unmanned vehicles. Lear Siegler Publication.

Lauck, K. E. 1986. New Standards for industrial robot safety. *CIM REVIEW—Robot Safety:* 60–68.

Rensselaer Polytechnic Institute. 1987. Center for Manufacturing Productivity and New Technology. *Robot Safety,* Final Report.

Trouteaud, R. R. 1979. Safety, training and maintenance: their influence of the success of your robot application. Report No. MS79-778. pp. 1–8. Dearborn, MI: SME.

Worn, I. H. 1980. Safety equipment for industrial robots. Robotics International Report No. MS80-714. pp. 1–13. Dearborn, MI: SME.

# 4

# Development of the ANSI/RIA Robot Safety Standard

James A. Peyton

## 1. INTRODUCTION

The Robotic Industries Association[1] (RIA) Robot Safety Standard was approved by the American National Standards Institute[2] (ANSI) on June 13, 1986. The revision to the landmark 1986 Robot Safety Standard, ANSI/RIA R15.06-1986, began in 1987. The R15 Executive Committee decided to review the landmark document within a year of its publication instead of the ANSI requirement for examination after 5 years. In September 1987, the Occupational Safety and Health Administration (OSHA) referenced the R15.06 standard in its *Guidelines for Robotic Safety* (OSHA, 1987). The OSHA guidelines were sent to all OSHA compliance officers. New OSHA guidelines based on the current revision of the 1986 standard are in preparation.

This chapter reviews the content and structure of ANSI/RIA R15.06. This chapter also outlines the proposed changes contained in Revision 10. Specific quotations from the draft revision are included.

The R15.06 Subcommittee review has seen 10 revisions of the standard since 1987. In 1989, the draft standard was sent to canvassees throughout the United States for review and comment. As of this writing, a second Canvass review of draft Revision 10 was completed. Publication of the revised standard is expected in the third quarter of 1991.

---

[1]The Robotic Industries Association is the only trade association in North America dedicated to the promotion of robotics. RIA activities include trade shows, seminars and conferences, standards development, and trade statistics.

[2]The American National Standard Institute, Inc. is the clearinghouse for standards developed in the United States. ANSI is also the official organization representing the United States in the international standards activities of the International Organization for Standardization (ISO) and the International Electrotechnical Commission (IEC).

## 2. STRUCTURE

The basic structure of the robot safety standard remains unchanged from the 1986 version. The three major sections (4, 5, and 6), cover manufacturer responsibilities, system requirements, and safeguarding respectively. Section 4—Manufacture, Remanufacture and Rebuild covers the requirements for robot construction. This section enumerates requirements for guarding against hazards to personnel associated with the design of a robot or robot system. Section 5—Installation of Robots and Robot Systems is the so-called "system" section. Section 5 gives design considerations for the installation of robots and robot systems. These design considerations include the hazards outlined in Section 4 of the standard.

Section 6—Safeguarding Personnel contains three main subsections. The first on responsibility includes the new sections on Risk Assessment and Stages of Development. The second subsection describes examples of and requirements for safeguarding devices. The last portion covers the safeguarding requirements for teaching, automatic operation, attended continuous operation, and maintenance and repair.

Sections 7, 8, and 9 cover maintenance, testing and start-up, and safety training of personnel, respectively. Section 7 does not go into great detail regarding maintenance schedules and required inspections, but does place the responsibility on the user to perform these operations on a regular basis. Section 8 provides a checklist for system test and start-up, including an exhaustive list of precautions for initial start-up. Section 9 on safety training provides general guidelines for training program content and retraining requirements.

The following proposed changes to the 1986 standard are included in Revision 10:

- Addition of illustrations.
- Expansion of Section 9—Safety Training of Personnel.
- Addition of an Application section.
- Addition of slow speed control.
- Addition of detection of failure to complete intended robot motion.
- Addition of a section on Risk Assessment.
- Addition of Attended Continuous Operation.

## 3. DEFINITIONS

The robot safety standard presents several definitions. The following definitions will aid in understanding this chapter.

**Attended Continuous Operation:** The time when robots are performing production tasks under slow speed control through attended program execution.

**Attended Program Verification:** The time when a person within the restricted envelope (space) verifies the robot's programmed tasks at programmed speed.

**Coordinated Straight Line Motion:** Control wherein the axes of the robot arrive at their respective end points simultaneously, giving a smooth appearance to the motion. Control wherein the motions of the axes are such that the tool center point (TCP) moves along a prespecified type of path (line, circle, or other).

**Emergency Stop:** The operation of a circuit using hardware-based components that overrides all other robot controls, removes drive power from the robot actuators, and causes all moving parts to stop.

**Enabling Device:** A manually operated device which when continuously activated permits motion. Releasing the device shall stop robot motion and motion of associated equipment that may present a hazard.

**Envelope (Space), Maximum:** The volume of space encompassing the maximum designed movements of all robot parts including the end effector, workpiece, and attachments.

**Industrial Robot:** A reprogrammable multifunctional manipulator designed to move material, parts, tools, or specialized devices through programmed motions for the performance of a variety of tasks.

**Industrial Robot System:** A system that includes industrial robots, the end effectors, and the devices and sensors required for the robots to be taught or programmed, or for the robots to perform the intended automatic operations, as well as the communication interfaces required for interlocking, sequencing, or monitoring the robots.

**Joint Motion:** A method for coordinating the movement of the joints such that all joints arrive at the desired location simultaneously.

**Operating Envelope (Space):** That portion of the restricted envelope (space) that is actually used by the robot while performing its programmed motions.

**Rebuild:** To restore the robot to the original specifications of the manufacturer to the extent possible.

**Remanufacture:** To upgrade or modify robots to the revised specifications of the manufacturer.

**Restricted Envelope (Space):** That portion of the maximum envelope (space) to which a robot is restricted by limiting devices. The maximum distance the robot can travel after the limiting device is actuated defines the boundaries of the restricted envelope (space) of the robot. [*Note:* The safeguarding interlocking logic and robot program may be such that the restricted envelope (space) is redefined as the robot performs its application program.]

**Safeguard:** A barrier guard, device, or safety procedure designed for the protection of personnel.

**Slow Speed Control:** A mode of robot motion control where the velocity of the robot is limited to allow persons sufficient time to either withdraw from hazardous motion or stop the robot.

**Start-up:** Routine application of drive power to the robot/robot system.

**Start-up, Initial:** Initial drive power application to the robot/robot system after one of the following events:

- Manufacture or modification.
- Installation or reinstallation.
- Programming or program editing.
- Maintenance or repair.

## 4. ADDITION OF ILLUSTRATIONS

One of the most frequent comments concerning the 1986 standard was that illustrations would help explain key concepts. The Subcommittee agreed to develop illustrations in three areas. The first set of illustrations shows robot maximum, restricted, and operating envelopes for different robot configurations. (See Figure 4–1.) The purpose of these illustrations is to demonstrate that the various robot envelopes may extend behind the robot as well as in front.

The second group of illustrations gives examples of safeguarding devices.

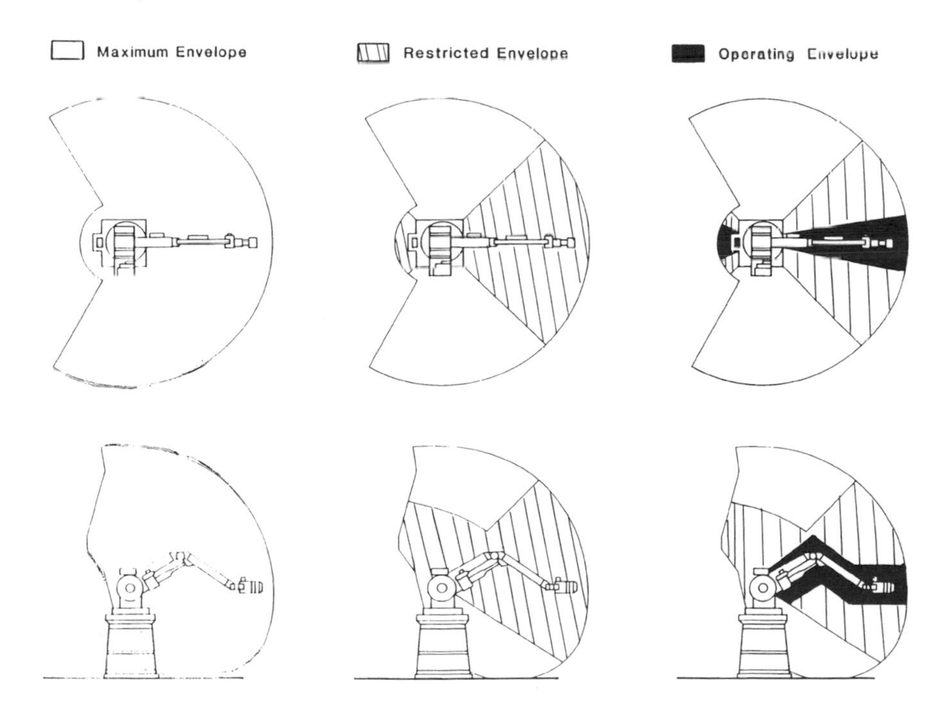


**FIGURE 4–1.** Maximum envelope, restricted envelope, and operating envelope for different robot configurations.

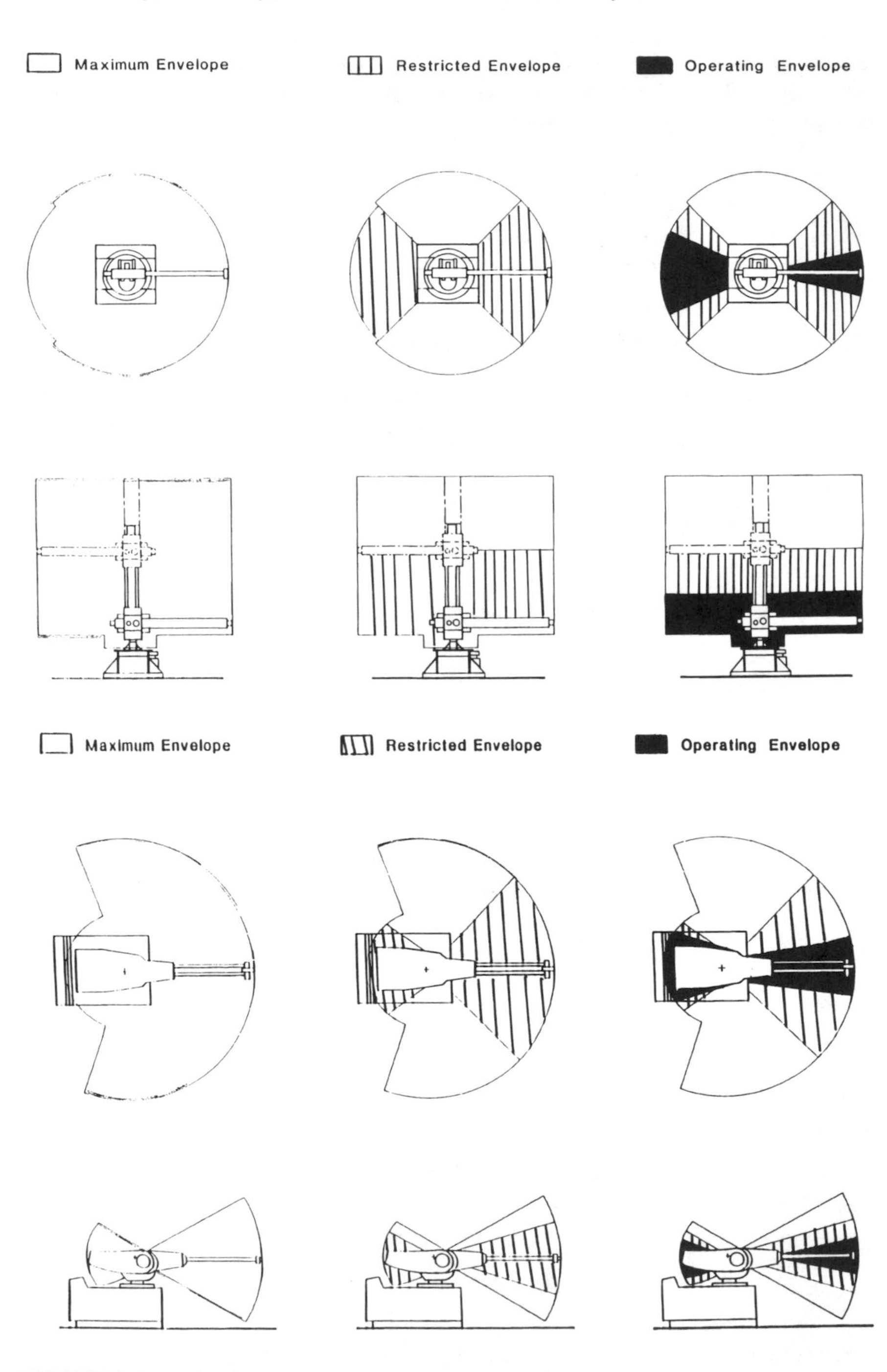


**FIGURE 4–1.** Continued

(See Figure 4–2.) The Subcommittee does not advocate a particular workcell configuration nor specific application of safeguarding devices. Development of safeguarding requirements is the result of a risk analysis as outlined in Section 6.1.1 of the standards.

The last illustration demonstrates a dynamic restricted envelope. (See Figure 4–3.) The restricted envelope changes allowing operator access to the deactivated portion of the workcell.

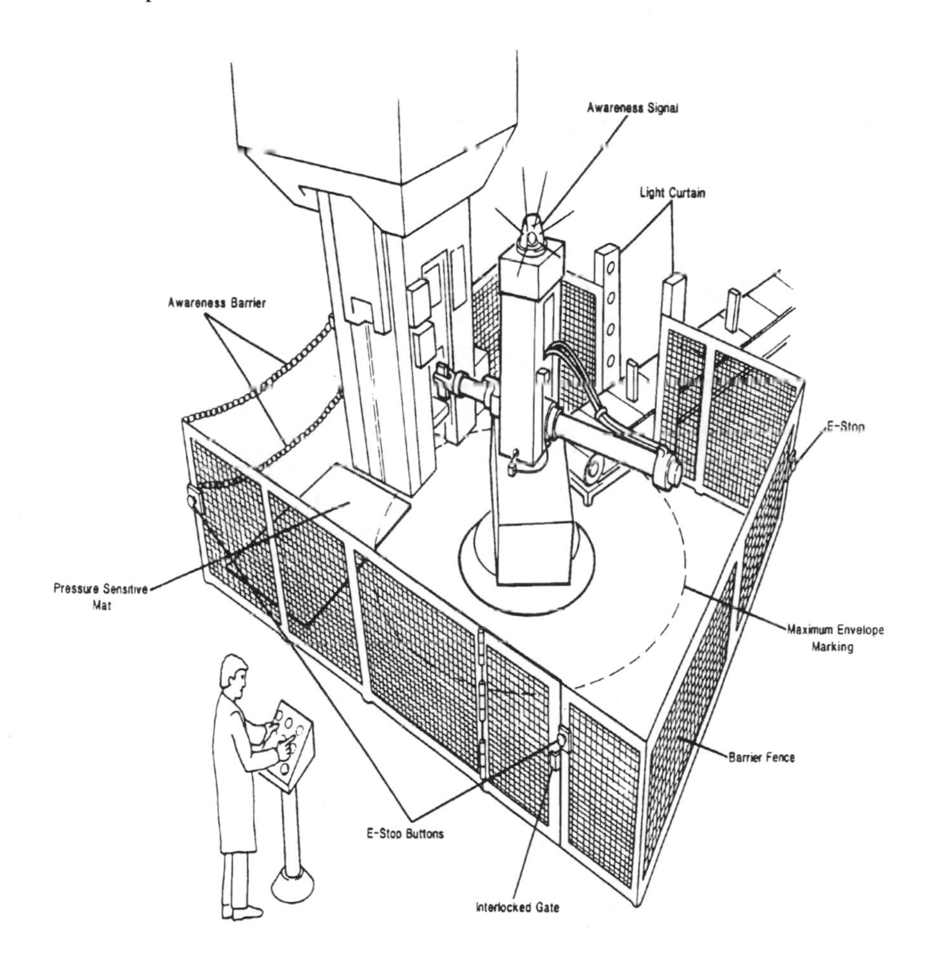


**FIGURE 4–2.** Illustration of robotic workcell with various safeguarding methods. (*Note:* This illustrates various methods of safeguarding requirements that shall be determined by the user in accordance with Section 6 of the standard.)

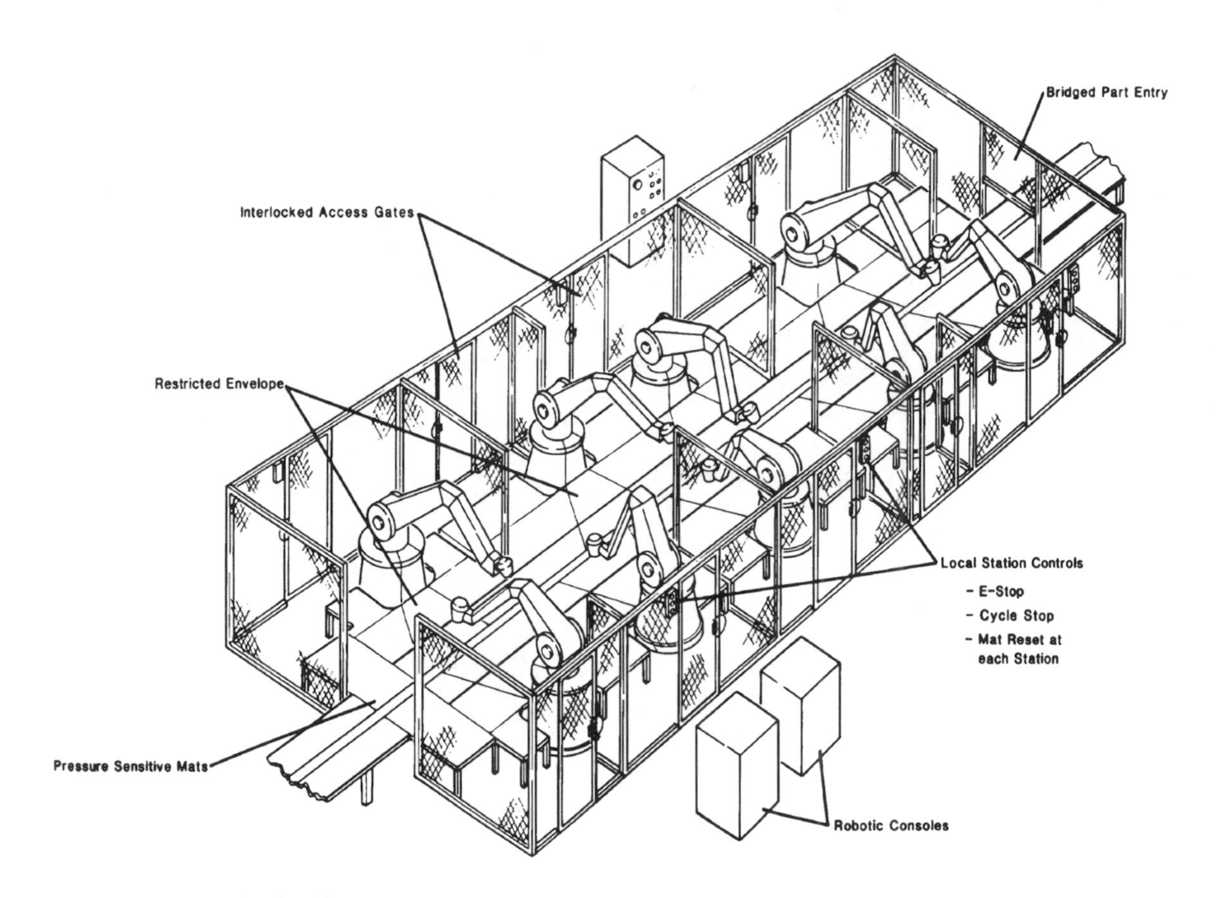


**FIGURE 4–2.** Continued

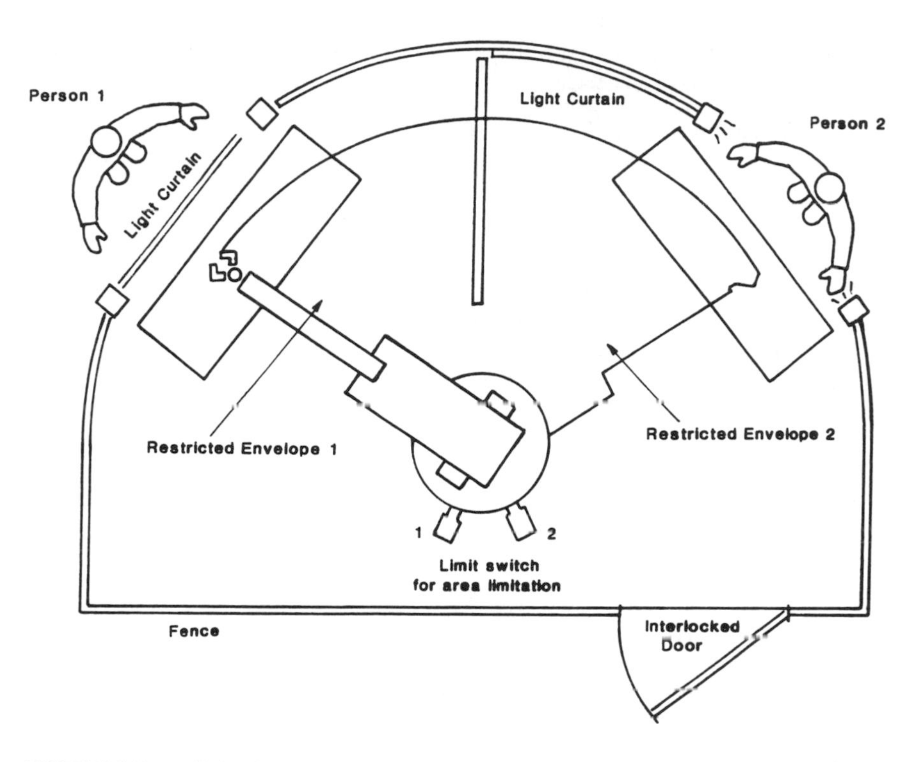


**FIGURE 4–3.** Robot installation with shifting work stations. The light curtains and the two limit switches (1 and 2) are so designed that the robot is stopped if it tries to enter an area with a light curtain beam broken.

## 5. ADDITION OF APPLICATION SECTION

Many industrial standards users are not aware that industry-generated standards can have the force of law. Under the General Duty Clause of the Occupational Safety and Health Act,[3] the employer must provide a safe workplace. In the absence of an OSHA regulation, accepted industry standards may be used as the basis for citations. The R15.06 Subcommittee developed guidelines for implementing the provisions of R15.06. Section 1.3 outlines the proposed application periods.

### 1.3 Application

**1.3.1 New Robots/Robot Systems.** The requirements of this standard shall apply to all newly manufactured robots and robot systems within eighteen (18) months after the approval date of this standard.

[3]Occupational Safety and Health Act of 1970 (29 U.S.C. 657, 667).

**1.3.2 Existing Robots.** The requirements of this standard shall apply to existing robots when remanufactured or rebuilt.

**1.3.3 Existing Robot Systems.** The requirements of this standard shall apply to existing robot systems when robots in the system are remanufactured or rebuilt.

Why add a compliance provision to a voluntary standard? Without the industry-generated compliance guidelines, the standard would be in force from the date of its publication. The proposed Application section allows a period for industry adjustment. Precedents may be found for this approach in machine tool standards, both published and under development.

## 6. SECTION 4—MANUFACTURE, REMANUFACTURE, AND REBUILD

Section 4 of the robot safety standard covers manufacturer requirements for industrial robots. Original equipment manufacturers, as well as remanufacturers and rebuilders must comply with this section. Section 4 enumerates several hazards to personnel to be eliminated by design or protection. These hazards include moving parts, component malfunction, and energy sources. There are also provisions for protecting against electromagnetic interference (EMI), radio-frequency interference (RFI), and electrostatic discharge (ESD).

Section 4 also provides requirements for actuating controls and Emergency Stop circuits. Actuating controls must be protected from unintended operation and must clearly indicate the robot status. Emergency Stops shall be hardware-based and be located at each operator control station, including control pendants. Provisions for additional Emergency Stops must also be included.

Section 4 requirements also contains guidelines for control pendants. Automatic operation cannot be initiated from the pendant exclusively. This stipulation prevents initiation of automatic operation from within the restricted envelope. For those using the pendant for attended program verification, the standard requires the pendant to have an enabling device (i.e., deadman's handle).

Other robot mechanical requirements include specifications for hard stops, provisions for lifting, movement without drive power, electrical connectors, and hoses. There are also criteria for preventing hazards in the event of power loss or general failures.

Two new Section 4 requirements cover detection of failure to complete intended motion and attended continuous operation. Another departure from the 1986 standard is slow speed control.

## 7. DETECTION OF FAILURE TO COMPLETE ROBOT MOTION

Detection of failure to complete intended robot motion is under consideration for addition to both Sections 4 and 5 of the standard. The purpose of this robot capability is to provide a safeguard to stop the robot in case someone is pinned between the robot and an immovable object. R15.06 received a specific request to develop language for this capability in response to two robot fatalities where the robot continued to press against the victim.

> **4.2.2 Detection of Failure to Complete Intended Robot Motion.** When a failure to complete intended robot motion presents hazards to personnel, the robot shall have a means of detecting a failure to complete intended motion. Detection of a failure to complete intended motion shall override other robot controls, remove drive power from robot actuators, cause moving parts to stop and provide an alarm output.

## 8. ADDITION OF ATTENDED CONTINUOUS OPERATION

Worker intervention in the restricted envelope has been debated since the revision began in 1987. RIA received comments from end users with welding applications where an operator inspected the weld as it was performed. The operator was able to inspect the weld and adjust the process using a pendant control. For obvious reasons this proposed addition has been the most controversial, as the numerous comments received during the first Canvass letter ballot indicate.

Sections 4.7 and 6.7 are designed to safeguard a person within the restricted envelope while the robot performs its programmed task under slow speed control. Both the robot requirements (Section 4.7) and the safeguarding requirements (Section 6.7) specify the use of a pendant that includes an enabling device.

> **4.7 Attended Continuous Operation.** Robots having attended continuous operation capability shall meet the following requirements:
>
> **4.7.1** Selection of attended continuous operation shall ensure that motion is under slow speed control as described in 4.8.
>
> **4.7.2** Selection of attended continuous operation mode shall only be possible from outside the restricted envelope (space).
>
> **4.7.3** Attended continuous operation shall require the continuous actuation of an enabling device by the operator. Releasing the enabling device shall remove drive power and provide an output signal through a hardware-based circuit.
>
> **4.7.4** Single point of control is required with attended continuous operation.

**6.7 Safeguarding During Attended Continuous Operation.** Certain robot applications may require one person to be present in or near the restricted envelope (space). The purpose of this section is to safeguard that person within the restricted envelope (space) while the robot is performing its programmed task under slow speed control. The person shall be safeguarded from the motion of adjacent robots or other associated industrial equipment that may present a hazard.

**6.7.1** The robot shall conform to Section 4.7.

**6.7.2** The pendant shall conform to Section 4.6.

In addition, the pendant shall have the following capabilities:

1. Release of an enabling device shall stop the robot motion and the operation of associated equipment that may create a hazard.
2. Initiation of pendant Emergency Stop shall override all other robot and associated equipment controls, remove drive power from robot and associated equipment actuators, and cause all functions to stop if continued operation may cause a hazard.

**6.7.3** When the attended continuous operation is selected, the following conditions shall be met:

1. The person shall have single point of control of the robot system.
2. When operating under drive power, slow speed control shall be in effect (see Section 4.8). The limits for slow speed control shall be selected such that there is sufficient time to either withdraw from hazardous motions or stop the robot.
3. All robot system Emergency Stop devices shall remain functional.
4. The person shall have sole control of movement of other equipment in the restricted envelope (space) if such movement would present a hazard.

**6.7.4** The person shall be trained in accordance with Section 9.

**6.7.5** To restore automatic operation the following shall be required:

1. Exit the restricted envelope (space).
2. Restore safeguards required for automatic operation.
3. Initiate a deliberate start-up procedure.

Obviously attended continuous operation is not universally acceptable. The Subcommittee expected to receive criticism during the initial letter ballot of the draft revision. The issue of operator intervention must be addressed if companies continue the practice despite current safety standard restrictions.

## 9. ADDITION SLOW SPEED CONTROL

The new section on slow speed control is a radical departure from the 1986 standard. Section 4.6 of ANSI/RIA R15.06-1986 states

> All robots shall have a slow speed. The maximum speed of any part of the robot shall not exceed 250 millimeters per second (10 inches per second). The robot shall be designed and constructed so that, in the event of any single, reasonably foreseeable malfunction, the speed of any individual axis shall not exceed 250 millimeters per second (10 inches per second). (ANSI, 1986)

In principle, the provisions of the 1986 standard are admirable. However, in reality these conditions are difficult, if not impossible, with which to comply. The R15.06 Subcommittee proposed the following:

> **4.8 Slow Speed Control.** All robots with pendants and robots that support attended continuous operation shall provide slow speed control. Settable slow speed velocity limits determine the maximum speed of the robot under slow speed control. These velocity limits shall be applied as follows:
>
> 1. When operating in coordinated straight line mode, the speed of the tool center point (TCP) shall not exceed 250 mm/sec (10 inches/sec).
> 2. When operating in joint mode, the maximum speed of each joint shall be reduced so that the speed of any part of the robot shall not exceed 250 mm/sec (10 inches/sec).
>
> The robot shall be designed and constructed so that in the event of any single reasonably foreseeable malfunction, robot speed shall not exceed these slow speed velocity limits.

The R15.06 proposal recognizes that different types of robot movement will result in different velocity limits. For example, in coordinated straight line motion, in order to maintain a 250 mm/sec speed at the tool center point (TCP), other parts of the robot may be moving much faster. In recognition of this fact, safeguarding sections referring to slow speed control have specific reminders that other parts of the robot may not be limited to 250 mm/sec.

## 10. SECTION 5—INSTALLATION OF ROBOTS AND ROBOT SYSTEMS

Section 5 is commonly referred to as the "systems section," for it is the application of Section 4 hazard prevention at the user facility. The installation requirements cover electrical grounding, power disconnect, and power sources. This section also covers clearance requirements and restricted envelope identification. System Emergency Stop specifications include provisions for multiple Emergency Stop locations.

Section 5 also covers the proposed section on system detection of failure to complete intended robot motion. This requirement is the application of Section 4.2.2 to the user facility. The proposed new section reads as follows:

**5.14 Robot System Detection of Failure To Complete Intended Robot Motion and Response**

**5.14.1** When the robot detects a failure to complete intended motion (See Section 4.2.2), the robot system and all other equipment shall stop and an alarm shall be activated if continued operation would present a hazard.

**5.14.2** Restarting the robot system shall require a deliberate action by the operator to follow a prescribed start-up procedure which shall take place outside of the restricted envelope (space).

## 11. SECTION 6—SAFEGUARDING PERSONNEL

The user has the ultimate responsibility for the safety of personnel associated with robots and robot systems. This section outlines requirements for safeguarding personnel in four activities: teaching, automatic operation, attended continuous operation, and maintenance and repair. Section 6 of the standard provides guidelines for risk assessment, usage of safeguarding devices, and specific requirements for the aforementioned activities.

One of the most important principles established in the safeguarding section is that any person in close proximity of the robot shall have single point of control. The safeguards used to protect personnel must be control reliable. The standard also indicates that where safeguards are defeated (e.g., during maintenance operations), alternate safeguarding is required to prevent injury.

## 12. ADDITION OF RISK ASSESSMENT

The 1986 safety standard alluded to an assessment of hazards to determine the level of safeguarding required. The R15.06 Subcommittee decided to borrow the language for a risk assessment section from the draft safety standard prepared by International Organization for Standardization (ISO), ISO/DP 10218 (ISO, 1989). A section on the stages of development was also added in recognition of the different safeguarding requirements at each stage. A robot being installed and programmed may require different safeguarding than one in production. The type and level of safeguarding are determined by the risk assessment.

**6.1.1 Risk Assessment.** The means of safeguarding shall be based on a risk assessment. The risk assessment shall consider:

1. The size, capacity and speed of the robot
2. The application/process
3. The anticipated tasks that will be required for continued operation
4. The hazards associated with each task

5. Anticipated failure modes
6. The probability of occurrence and probable severity of injury
7. The level of expertise of exposed personnel and the frequency of exposure.

**6.1.2 Stages of Development.** Safeguarding requirements also vary for the different stages of development of robots and robot systems. The following stages of robot and robot system development are recognized:

1. Integration at the manufacturer or system developer
2. Verification and buy-off testing
3. Installation and testing on site of operation
4. Operation in production

These stages involve personnel of different levels of skill and require different safeguarding. For example, during integration at the manufacturer, any elaborate safeguarding is likely to impede the development of the system making it impractical to debug; while for operation in production, full safeguarding and interlocks may be a requirement for the same system. The level of personnel safeguarding shall be applied at each stage consistent with applicable risk assessments.

## 13. SECTION 6.3—SAFEGUARDING DEVICES

Section 6.3 establishes performance requirements for safeguarding devices. Three main categories of safeguarding devices are presented: limiting devices, presence-sensing safeguarding devices, and barriers.

Limiting devices are further categorized as either mechanical limiting devices (e.g., pins) or nonmechanical limiting devices (e.g., limit switches, relays, pull plugs). Both types of limiting devices must be capable of stopping the robot at its rated speed and load. Nonmechanical limiting devices must be designed, constructed, and installed such that any single component failure, including output devices, shall not prevent the normal stopping action of the robot.

Presence-sensing safeguarding devices include light curtains, pressure-sensitive mats, and proximity detectors. The standard requires that these devices be designed and installed so that ambient factors will not affect normal operation. If a presence-sensing device is actuated, the standard establishes procedures for resuming robot operation. The devices must be designed and installed so that component failure will not affect the normal stopping action of the robot. In addition, automatic operation will be prevented until the failure is corrected.

Barriers such as fences must prevent personnel from reaching the restricted envelope. If access through a barrier is necessary, the standard requires the use of an interlocked gate. The interlocked access must either stop the robot and remove drive power to the robot actuators, or stop automatic operation of the robot and any other associated equipment that may cause a hazard. Once the interlocked access is opened, to restore automatic operation the following is required:

1. Exit the restricted envelope.
2. Restore safeguards required for automatic operation.
3. Initiate a deliberate start-up procedure.

Interlock mechanisms must be designed so that any component failure will not affect the stopping action of the robot. Automatic operation will be prevented until the fault is corrected.

## 14. SAFEGUARDING THE TEACHER, OPERATOR, AND OTHER PERSONNEL

Sections 6.5 through 6.8 of the standard cover safeguarding requirements for specific robotic activities: teaching, automatic operation, attended continuous operation, and maintenance and repair. Each section requires that persons be trained in accordance with Section 9 **Safety Training of Personnel.** Each robotic operation requires that persons controlling the robot must have single point of control. Both Section 6.5 **Safeguarding the Teacher** and Section 6.7 **Safeguarding During Attended Continuous Operation** provide for robot motion under slow speed control. Each category of robotic operation defines methods for initiating robot motion and how to restart robot motion if the robot is shut down.

## 15. MAINTENANCE, TESTING, AND START-UP

Section 7 **Maintenance** and Section 8 **Testing and Start-up** provide guidelines for maintaining the robot in good working order. Section 7 requires that robots receive periodic maintenance. Section 8 specifies the method for function-testing robots and robot safeguarding devices before start-up. The standard differentiates between a routine application of drive power and start-up after maintenance, program change, or repair. The latter, known as "initial start-up," requires the user to function test the components of the robot system outside the restricted envelope and verify that all is in good working order.

## 16. EXPANSION OF SECTION 9—SAFETY TRAINING

Another frequent comment concerning the 1986 standard was the lack of training guidelines. The Subcommittee added sections on Training Program Content and Retraining requirements. It should be noted that exact training requirements are application specific. The user is responsible for the development of a training program to ensure that any person who programs, teaches, operates, maintains, or repairs robots or robot systems is trained and demonstrates competence to perform the assigned task safely.

The new provisions for training program content are as follows:

**9.2 Training Program Content.** Training as appropriate to each assigned task should include, but not be limited to:

**9.2.1** A review of applicable industry safety procedures and standards such as BSR/RIA R15.06-19XX.

**9.2.2** A review of applicable robot vendor safety recommendations.

**9.2.3** An explanation of the purpose of the robot system and its operation.

**9.2.4** An explanation of the specific tasks and responsibilities for each person.

**9.2.5** The person or persons (by name, location and phone number) to contact when the actions when the actions required are beyond the training and responsibility of the person being trained.

**9.2.6** Identification of the recognized hazards associated with each task.

**9.2.7** Identification of, and appropriate responses to, unusual operating conditions.

**9.2.8** An explanation for function testing or otherwise assuring the proper functioning of safeguarding devices.

## 17. REQUEST FOR INPUT

The RIA Robot Safety Standard is a consensus document. It was developed by industry in order to provide guidelines for the safe application of industrial robots. Just as the 1986 standard was reviewed shortly after its completion, the current revision will be examined carefully after its publication as an ANSI standard. No standards development activity can be successful without the input from a wide variety of interests. RIA welcomes input to future revisions of the robot safety standard. Specific questions or comments should be addressed to the Robotic Industries Association, P.O. Box 3724, Ann Arbor, MI 48106.

**References**

ANSI. 1986. ANSI/RIA R15.06-1986. *American National Standard for Industrial Robots and Robot Systems—Safety Requirements*. New York: American National Standards Institute.

ISO. 1989. ISO/DP 10218 *Manipulating Industrial Robots—Safety*. Geneva: International Organization for Standardization. *(For more information regarding ISO activities contact RIA c/o USA TAG/ISO TC184/SC2.)*

OSHA. 1987. *Guidelines for Robotic Safety,* OSHA Instruction PUB 8-1.3. Washington, DC: OSHA.

RIA. 1990. BSR/RIA 15.06. *Proposed American National Standard for Industrial Robots and Robot Systems—Safety Requirements (Rev. 10)*. Ann Arbor, MI: Robotic Industries Association.

# 5

# Reliability and Safety in Teleoperation

Robert H. Sturges, Jr.

## 1. INTRODUCTION

The teleoperator as a robotic device precedes the industrial robot historically by several decades. The first commercial mechanical remote manipulator devices (Figure 5–1) were employed in research and defense production at the Argonne National Laboratories in 1948 (Goertz, 1949). These devices transferred, in pantograph style, the forces and displacements applied by the operator at the "master" end to the remote "slave" end. These robust and well-engineered devices featured very low friction and inertia, and naturally offered bilateral force reflection. By their design and application, the distance between the master and slave was limited; therefore, line-of-sight control (typically through leaded glass or with mirrors) was required. Advanced models of these early devices replaced the direct mechanical connection of master to slave with servos (Figure 5–2) as early as 1954 (Goertz, 1954).

We recognize two basic forms of teleoperator today: the unilateral and the bilateral. The unilateral teleoperator provides remote controls with which force and motion can be transmitted from the operator controls to the actuators. Feedback is provided by a vision system, usually closed-circuit television (CCTV), and sensors, which monitor remote conditions and display them to the operator. The bilateral teleoperator features "forward" control as the unilateral type, but also permits "reverse" control in which force and motion can be transmitted by the remote actuators to a similar set of local actuators controlled by the operator. This innovation provides "feel" for the operator in addition to the other feedback. The term "master/slave" applies to either unilateral or bilateral types in which motions (and forces) are proportionally reproduced from the controls (master) to the actuators (slave). With the advent of on-line ma-

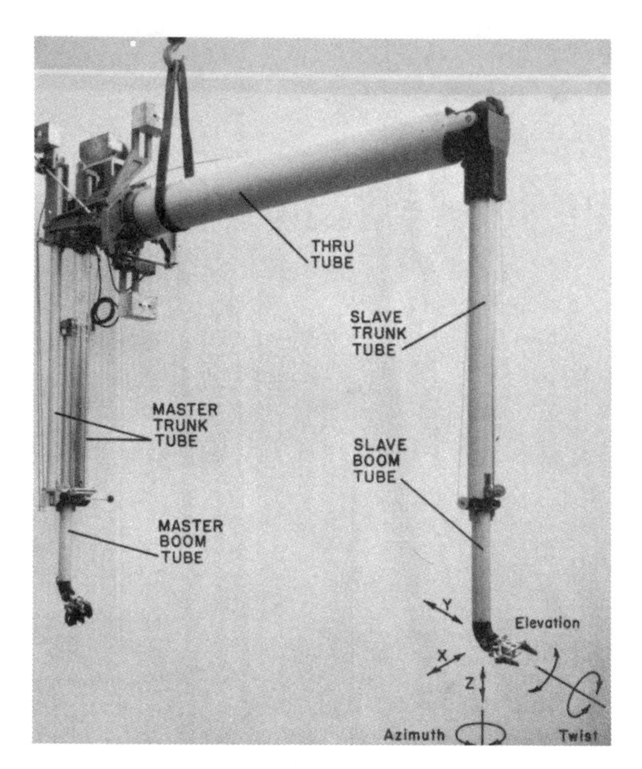


**FIGURE 5–1.** CRL Mod 8 photo from TC pg 3, (Corliss and Johnsen, 1969).

nipulator kinematic computations in real time, the physical configuration of the master and slave need not be similar to each other. The development of proportional electrohydraulic servo valves (Lee and Shearer, 1955) brought high power with precise control to the remote slave actuator. These devices expanded operating bandwidths beyond that needed by a human "master" operator.

These developments expanded the role of the electromechanical teleoperator into four application areas: manipulators, prosthetic and orthotic devices, walking machines, and human-amplifiers (Corliss and Johnsen, 1969). Each of these includes mechanical analogs of human motion, although they most often include far fewer degrees of freedom than found in nature. In addition, the functions of arms and legs may be embodied in sliding (prismatic) joints unlike those found in arthropods and vertebrate animals.

Manipulators may be fixed or attached to vehicles. They find use in hostile environments such as radioactive "hot cells," undersea maintenance and retriev-

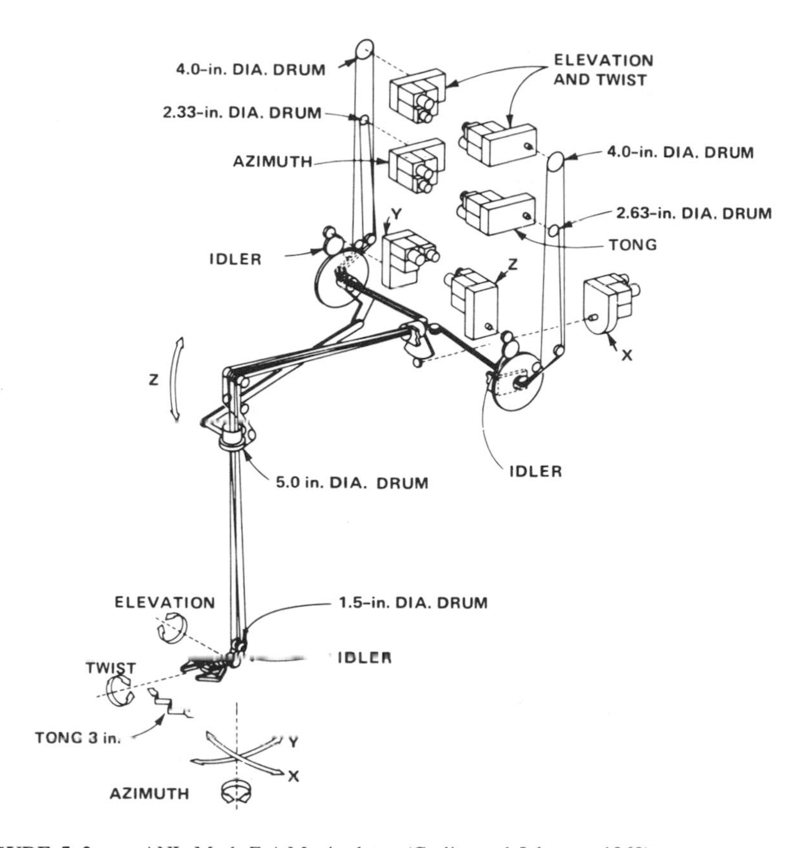


**FIGURE 5–2.** ANL Mark E-4 Manipulator (Corliss and Johnsen, 1969).

al, and, most recently, sampling planetary surfaces. The role of the "master" arm may be relegated to a programmable controller for positioning tasks requiring little or no feedback in a well-structured environment. Indeed, in this mode the manipulator serves as an industrial robot and shares the same reliability and safety issues. The temporary loss of function by a manipulator tends to be far more critical to its mission than that of the industrial robot, however, due to the greater difficulty (or impossibility) of repair or recovery in the remote or hostile environment. It is a common observation that otherwise well-trained operators of hot-cell manipulators chose to grasp even expendable and innocuous objects as sponges with a viselike grip to avoid dropping them. Research in manipulator "telepresence" is an active field (Tesar, 1989).

Prosthetic and orthotic devices attach directly to the human body and provide replacements for natural limbs or assistance to weakened members. The reliability of these devices depends strongly on the human–machine interface, which must take into account the dimensions and comfort of the individual user. Moreover, the motions and forces developed by prosthetic and orthotic devices directly affect the human user. Unlike the industrial robot or manipulator that operates in a hostile environment with no one in the immediate reach and workspace of the mechanisms, the user is within intimate "reach" of his or her robotic system at all times and needs to be protected from any unexpected or unreliable modes of motion. The forces needed to accomplish ordinary tasks can inflict serious injury to the user or the people around him or her if even momentarily misdirected.

The walking machine has evolved from simple cam-actuated mechanisms to multifreedom servomechanisms under sophisticated adaptive controls (Raibert et al., 1983; Song and Waldron, 1988). Walking machines for extraterrestrial exploration are currently the objects of intensive research (Kanade et al., 1989). When developed for practical use, the safety issues for such machines will be similar to conventional "off road" construction vehicles, viz., unstable ground contact conditions, steep angles of ascent and descent, and overall field reliability.

The human-amplifier concept enjoyed a brief period of research and development in the 1960s, beginning with a bilateral-force-reflecting quadruped walking machine, a (stationary) pair of anthropomorphic arms, and a full exoskeletal human-amplifier (Mosher, 1967). These high-powered hydraulic master/slave devices were unquestionably as dangerous as today's industrial robots. In addition, the dangers to the user of any force reflecting servo also apply to human-amplifiers, since he or she works among the powered axes at all times. A unique training and safety issue arises with these machines: the scaling of force with respect to mass has no basis in operator experience. Weights that are easily lifted by the powered exoskeleton can cause the entire machine to tip over due to the shifted center of mass of the ensemble. The intuition of the operator based on his or her "normal" operating stance would need to be modified to prevent mishap. Thus, issues of fatigue, perception, and judgment would need to be addressed in evaluating the failure modes of such systems.

The primary reliability concerns with respect to mission completion for each of the above manipulator systems are similar to those already discussed in Chapter 1 with regard to robot availability and service life. In addition, the presence of the human in the loop and the possibility of added control modes raises other issues, which include system integrity, data integrity, control design and task requirements. In the following sections we will introduce these safety areas and discuss their solutions. We will illustrate these with examples of working systems from industry and university research.

## 2. SYSTEM INTEGRITY

The safety of a teleoperator system can be measured with respect to two fundamental interests: mission safety and operator safety. When a mission involves danger to people (or animals), we may elect to refer to system reliability and safety as the same concern. When a mission involves only machinery (however expensive), we tend to refer only to reliability and mission survival. Operator safety becomes important in bilateral systems in which forces and motions routinely fed back to the user may take on values larger than he or she can safely respond to. Both of these concerns involve unreliable operation of the system in normal use, but may also involve the *reliable* performance of the system to abnormal disturbances. The possibility of people entering the workspace of a powered industrial robot is viewed as a safety issue, even when the system is otherwise operating reliably. The operator of a bilateral teleoperator master may be endangered when the slave is moved by an unexpectedly fast and powerful force, e.g., something dropping on it or slamming into it.

In this section we will limit our discussion to issues of system reliability, that is, potential causes of user/mission danger due to a loss of normal system integrity. In Section 4 we will discuss safety with respect to reliable but abnormal system operation.

### 2–1. Loss of Function

A teleoperator system may experience a loss of function leading to user or mission danger as a result of at least three causes: structural, power, or signal failure. Structural failure comprises loss of end-effector position due to breakage of a tendon (cable or tape), a structural member (arm, link, or pin), fastener, or a power transmission device (gear, rack, lead screw, key, shear pin, etc.) These occurrences are most often due to mechanical overload, which results from inadvertent strikes of the master or the slave with rigid objects, but can also be induced from loss of lubricant (in transmission devices) or stress corrosion cracking of initially sound material sections. Fastener failure has been observed in teleoperator slaves and traced to improper torquing during on-site assembly. For example, the portable hydraulic slave manipulator shown in Figure 5–3 is designed for breakdown into four sections for manual cartage to and from the remote work site. Each 35-pound section is joined rigidly with a single bolt and self-centering dovetail clamps. Bolt torque specifications are narrow, since insufficient torque would permit separation of the sections under end-effector-induced vibration, and over-torquing could result in fastener damage leading to expensive field repair.

Structural failure that does not render the system inoperable presents dif-

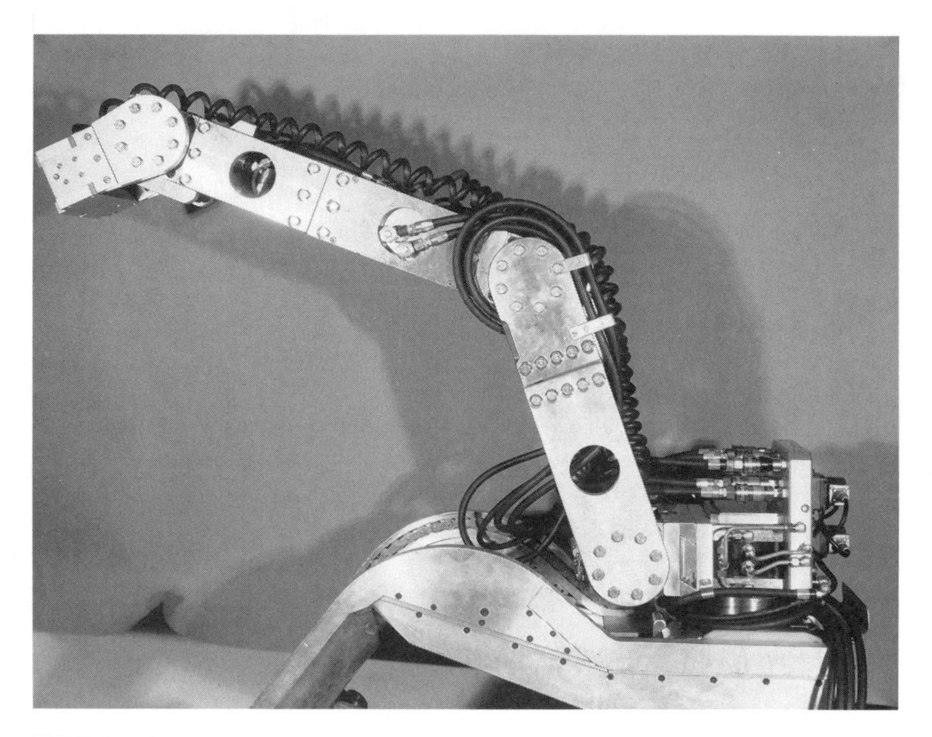

**FIGURE 5–3.** "SAM," a portable hydraulic slave manipulator.

ficulties of a more subtle kind. We will discuss these in the following section.

Power loss failures include the momentary or sustained inability of the actuators to deliver the forces commanded of them. Signal loss failures include the intermittent or continuous loss of actuator control or feedback signals, as distinguished from noise, which lowers the integrity of the signal itself. Several approaches can be applied to ameliorating these faults, viz., detection, redundancy, and override. Each of these applies to the specific power application domain of the teleoperator.

### *Detection*

Detection of power loss can be implemented at several system points. In electrically actuated robots, power line monitors can detect supply voltage loss within a fraction of a cycle, that is, long before the capacitors in a DC motor drive supply have begun to drain. Hydraulic systems similarly would be fitted with supply pressure sensors or pump power monitors for early detection of incipient pressure loss. AC servos under these conditions would feel the loss more quickly if tied directly to the incoming power line without synchronous condensers or UPS protection.[1] Supply power loss detection provides the control system with the opportunity to initiate a shutdown procedure. Such a procedure may need to consider the current system state. The definition of "fail safe" may indeed vary in the workspace and with the task being performed. For example, a loss of power condition occurring when the teleoperator is holding a sensitive piece of equipment or is positioned above such equipment requires that each joint hold its current position until power is restored. If power cannot be restored, each joint may need to be released "slowly" and in a particular order to minimize damage to the robot and its environment. This strategy may not be required in the zero gravity of space and undersea teleoperation. We will need to revise our definitions of "fail safe" continuously with each fault source and condition. For a more detailed discussion of failure modes and effects analysis (FMEA), see Jakuba (1986).

Hydraulic power fluctuation faults can be minimized or eliminated by considering each power path as a transmission line and providing local accumulators (by-passing), and reducing inductance (inertance) by system design and simulation (Blackburn et al., 1960; Paynter, 1953). Hydraulic systems require periodic purging; back pressure needs to be modeled and maintained to avoid the occurrence of cavitation.

Detection of signal loss can be implemented at several system points. Detection of signal loss can be implemented indirectly by monitoring the loop error in each axis controller. A sudden change or erratic series of values in axis error indicate a failure in the feedback system if they occur at rates in excess of the time constants of the intact physical system. Similarly, position error that exceed a specified limit indicate either excessive forces (you hit something and don't know it) or unresponsive actuators. Such detection schemes are common in numerically controlled (NC) equipment and industrial robots that are primarily position controlled. This approach is generally not viable in bilateral force/compliance-controlled applications in which the master and slave are expected to differ in their positions as a means of creating contact forces.

***Redundancy***

Teleoperators for hazardous or remote environments are often designed with redundant degrees of freedom, duplicate actuators, or duplicate members. For example, the hydraulic Naval Anthropomorphic Teleoperator (NAT) features nine degrees of freedom, imitating the flexibility of the human arm to negotiate obstructed workspaces. Loss of a single joint does not ordinarily render the unit incapable of useful work and permits its removal for repair.

[1]UPS (uninterruptable power supply) is a system comprising storage batteries, charging system, and DC-to-AC power inversion. Depending on battery energy capacity, an UPS may carry a system over power outages of a few seconds to several hours.

Teleoperators with nonredundant freedoms can be made resistant to single-point actuator failure through multiple motors that drive a common shaft or single motors with multiple windings (Greenaway and Sturges, 1979). In these designs, the duplicate motors normally operate in tandem with adequate power for all normal tasks. In the event of a single motor failure, the actuator set is reduced to half power and executes a modified mission profile until repaired.

The inclusion of duplicate members has been applied in hot cells since their inception. The loss of a single station in a row of manipulators requires changing the movement of materials and equipment in the cell with minor effect on the productivity of the system. Duplicate members are employed in the CMU/NASA Mars Rover (Kanade et al., 1989), a six-legged walking machine for planetary exploration (Figure 5–4). The lack of repair capability in extraterrestrial teleoperators requires novel approaches to reliability and safety to the mission. Its unique design permits the failure of one or two legs, while retaining the ability to walk. The inoperable legs are stowed out of the path of the other operable legs. An FMEA is used to guarantee the availability of this mode of failure recovery.

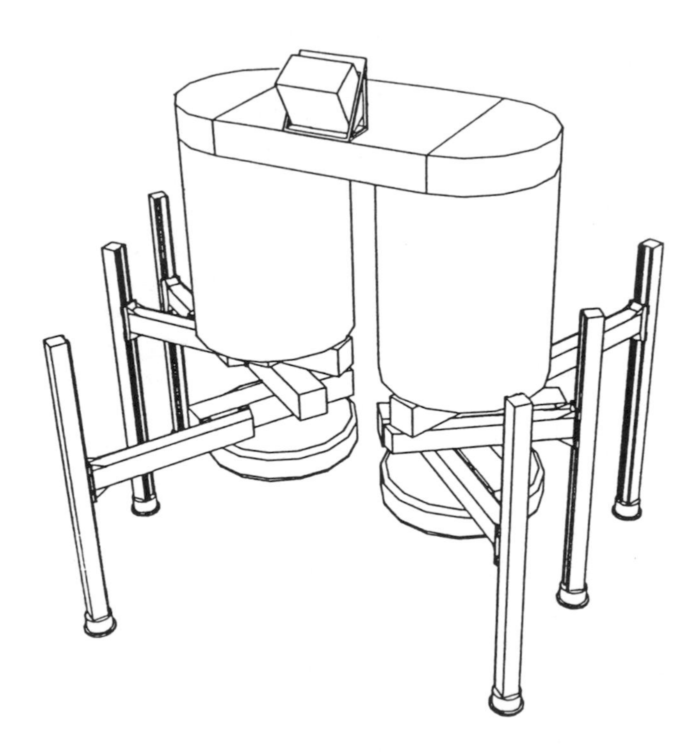

**FIGURE 5–4.** The "Ambler," Mars Rover, a six-legged walking machine for planetary exploration.

***Override***

For electrically actuated systems in which the primary action is maintaining position after a power loss fault, the addition of a shaft brake that holds upon loss of power (e.g., spring powered) should be considered. Modulating such a device after fault diagnosis can accomplish the gradual release of the affected axes. Shaft brakes that feature ratchets or dogs, rather than friction surfaces, provide positive position control with low stored energy and weight penalty. These devices require external assist (e.g., a crane or jack) to reposition the teleoperator's joint. Axis drives that cannot be back-driven (e.g., high-ratio worm gearing) provide this self-locking feature, but have the disadvantage of requiring an alternative mechanical override, such as access to the actuator shaft with a manual or power-driven wrench. In this case, the redundancy is not built in, but applied only when needed in extreme circumstances.

## 2–1. Reductions of Function

A teleoperator system may experience a structural reduction of function leading to user or mission danger that can be classified as follows: (i) temporary, stress induced; (ii) permanent, strain induced; and (iii) intermittent, kinematic lost motion. Reductions of function arise from the same sources as loss of function, but are more subtle, and in fact may pass undetected after their occurrence. In this section, causes and solutions of structural loss of function will be discussed. Section 3 will be devoted to signal and data-integrity issues.

Temporary, stress-induced failures are those that degrade the performance of the teleoperator during the time of application of a certain load, but are otherwise not present. For example, an incremental lift motion commanded from a master arm under no-load conditions may not produce the same motion under load. The relationship between commanded position and actual may indeed be highly nonlinear and dependent on the configuration of the slave joints. In a master/slave control scenario, this may not be a problem. However, if a record/playback mode is invoked after teaching a sequence of moves, the slave may respond differently to each of its loads. The weight of the load is not the only factor that needs to be considered in this case. The distribution of the mass in the load (i.e., its center of gravity and angular momentum) has a measurable and often large effect on end-effector position due to teleoperator joint and structural compliance (Menq, 1988).

End-point-position accuracy can be maintained for all loads and tasks when a model of the teleoperator is embodied in an operating simulator. Such a simulation would include the kinematic and dynamic parameters of each joint and actuator, and the kinematic and inertial characteristics of each expected load. Kinematic deflections are the easiest to calculate (Menq, 1988), and in many cases the deflections of the manipulator members may be lumped with the actuator deflections. The result of such simulations would be predictions of the actual end points of a manipulation task and the offsets needed of the master path program to compensate for teleoperator arm deflections. It is within the capability of state-of-the-art personal computers to perform such computations in real time (Book, 1987). Dynamic deflection compensation is still a research issue. Computation of manipulator dynamics without deflecting members is just becoming possible for six-degree-of-freedom systems with the recursive Newton–Euler method (Sahar and Hollerbach, 1986). This implies that accurate control of the robot *path* under varying load is not yet generally possible in real time.

Permanent, strain-induced failures are those that degrade the performance of the teleoperator during the time of application of a certain load, and remain present to some degree afterward. These usually occur because of mechanical overload due to inadvertent strikes of the master or the slave with rigid objects, but can also be induced from force and moment overloads of the actuators themselves. While it is rare that an animal can bend or break its own bones by muscle exertion, it is often possible that a manipulator device can strain itself to structural failure by inappropriate control commands.

Any manipulator link is comprised of four kinematic parameters: rotation about its axis, translation along its axis, translation normal to its axis, and rotation normal to its axis (as shown in Figure 5–5) (Paul, 1981). Thus, a six-degree-of-freedom teleoperator is modeled with 24 independent parameters, which manifest themselves in a highly nonlinear way at the end point. A strain in

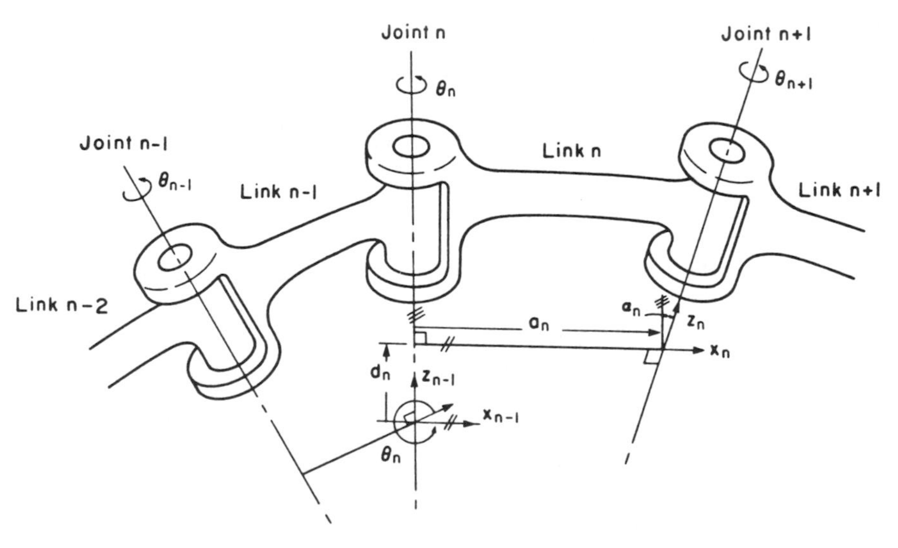


**FIGURE 5–5.** Manipulator link kinematic parameters. (Paul, 1981).

any one of them has varying consequences for end-point-position accuracy. Such models are now built into the most sophisticated robot controllers at the factory. They seldom retain their validity in use. Most rely on the "fact" that certain joints are precisely orthogonal to others, so that the kinematic equations that model them contain many "zero" values. Strains encountered in use often induce small angle changes that produce significant accuracy errors which no amount of "tweaking" the few parameters provided can remove.

Means are being devised to update these parameters using full parameter models and automatic on-line measurement of end-point positions (Stone, 1986), but commercial applications have not yet developed. Maintaining end-point accuracy by structural means remains a problem for general robotic devices.

Intermittent kinematic loss of accuracy comprises loss of end-effector position due to lost motion (play) in a tendon (cable or tape), a structural member (link or pin), a fastener, or a power transmission device (gear, rack, lead screw, key, shear pin, etc.). Such inaccuracies are highly dependent on the relative position of the load and the present arm configuration. Lost motion is almost always manifested about the actuator axis of motion, although the example in Section 2.1 cites a case in which the structure itself could assume an unexpected shape owing to inadequate fastener torque.

## 3. SIGNAL AND DATA INTEGRITY

Safe and reliable operation of an electromechanical teleoperator system relies on accurate and dependable data sources. The literature abounds with data-transmission and error-detection techniques in both the analog and digital domains (Cooling, 1986). This section will be concerned with techniques that ensure reliable data before the data enter the transmission medium, and how such data may be corrupted due to other factors particular to teleoperation. Three cases will be discussed: device degradation, time delay, and world model shifting.

### 3–1. Device Degradation

Signal loss detection as described above cannot guarantee that the data being transmitted between the master and slave is error free when the measurement devices themselves undergo degradation. Unfortunately, there is no universally applicable approach to this problem for all such feedback and control elements. However, a highly successful method employed for degradation detection in synchro/resolvers, a device class with wide use in the teleoperator and robotics field, will be illustrated.

Detection of signal loss can be implemented at several system points. In electrically controlled robots, cable connectors invariably provide the largest class of damage/loss sites. Signal integrity can be checked by wiring several of the connector pins in each cabel connector into a circuit that determines simple continuity in these wires (Figure 5–6), that is, zero or infinite impedance. Such a

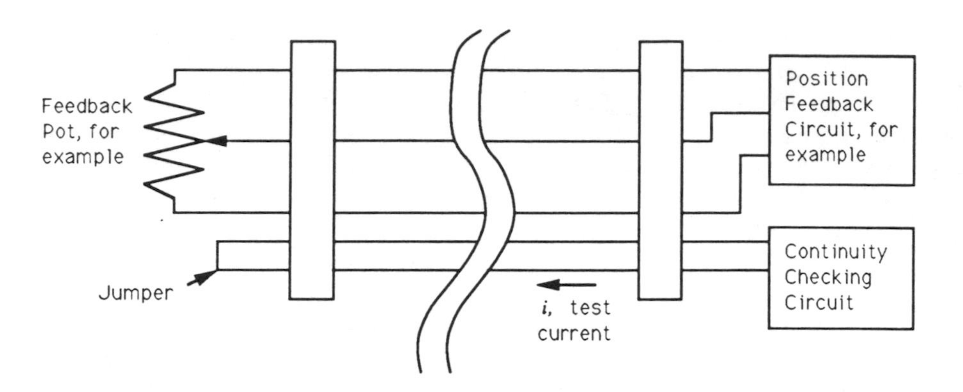


**FIGURE 5–6.** Signal integrity checking by continuity.

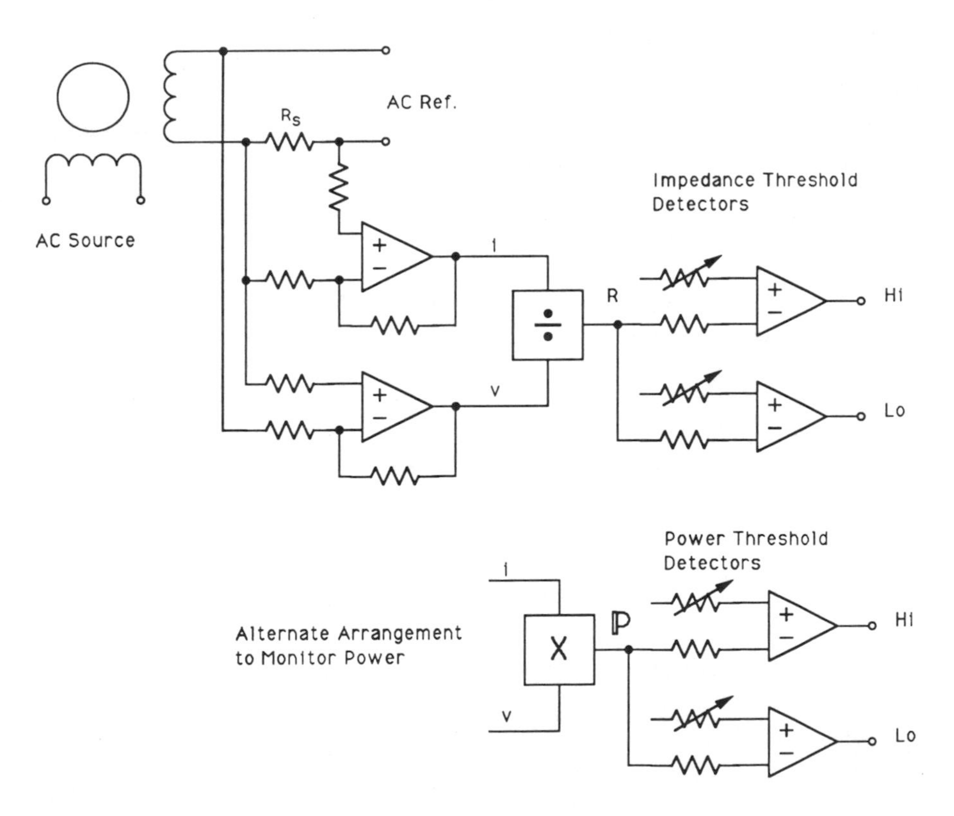


**FIGURE 5–7.** Impedance monitoring of synchro resolvers in the M/STI-SAM system.

detector, as mentioned in Section 2.1, operates independently of the control signals themselves, is easy to implement, and is highly immune to noise. It is incapable of detecting a fault in the control wiring itself, such as would occur if *only* a signal wire were open or shorted, or if a single control connector pin were faulty.

Signal integrity can also be checked by monitoring the impedance of the circuit elements *in situ*. Information can be derived from both quasistatic signal levels and higher frequency noise that may be imposed on it (Figure 5–7). For quasistatic monitoring with fixed thresholds, low-pass filtering is recommended for power and impedance signals (Figure 5–8), since they periodically drop to zero or saturate during the zero crossings of their inputs. High-pass filtering with the cutoff above the reference frequency may also indicate device failure owing to intermittent brush contacts or connector wiring.

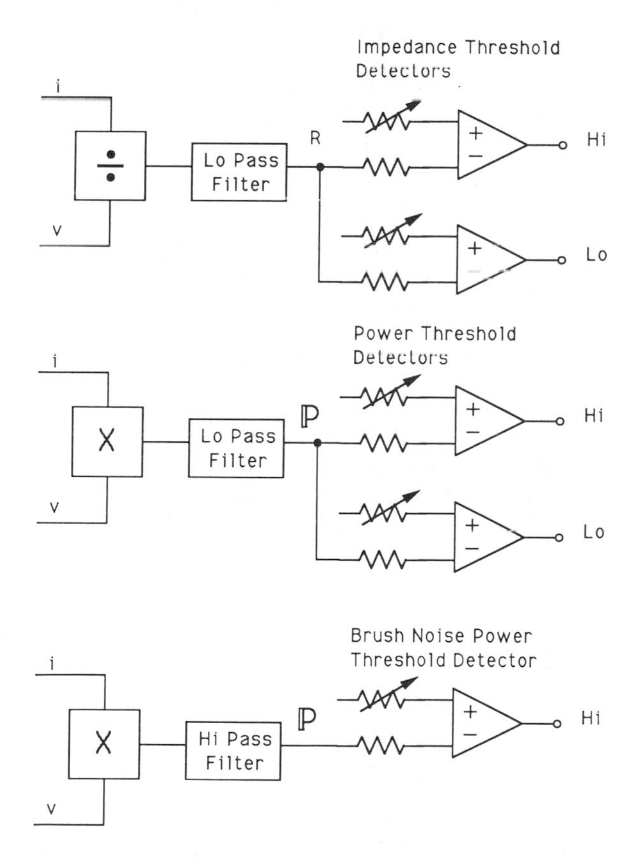


**FIGURE 5–8.** Noise and signal separation through filters.

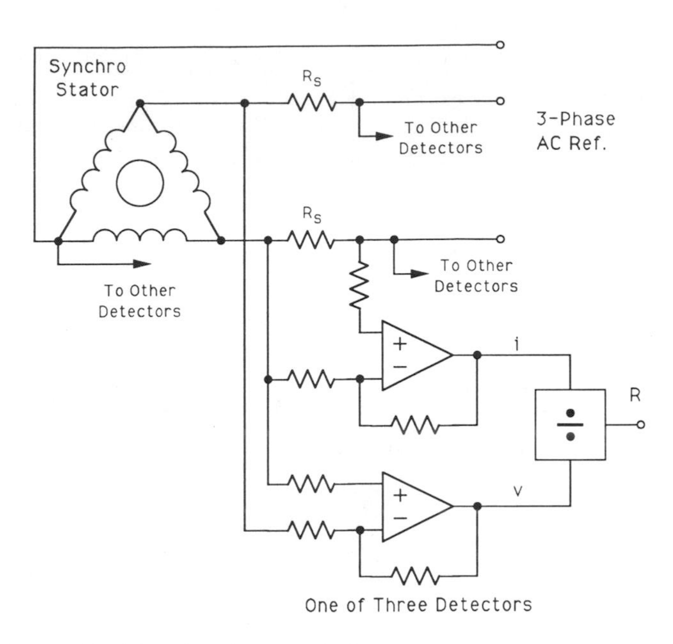


**FIGURE 5–9.** Multiphase impedance monitoring.

The stator impedances of synchro transmitter/receivers can be similarly monitored with additional circuitry. One set of voltage and current detectors per phase would be used in this case as shown in Figure 5–9.

## 3–2. Time Delay

Data passed to and from the remote location and the user of a teleoperator may be free from corruption due to transmission errors and feedback/display device characteristics and yet suffer from information loss in a practical sense from time delay. We have all experienced the effects of time delay in signal transmission: we are warned that the stopping distance of a moving car depends on the half-second or so that occurs between the time we note danger and the time our feet actually moves to apply the brakes. Such propagation delay limits the response times of closed-loop systems. Time delay in teleoperator systems has two sources: transmission delay and processing delay. Transmission delay arises from the fact that signals can be propagated almost at the speed of light. This delay is negligible for ordinary terrestrial applications, but for orbital, lunar, and planetary-based teleoperators, this delay is significant. For example, a geosynchronous satellite orbiting at a distance of 20,000 km experiences a total (round trip) propagation delay of 133 msec. As time delays increase, users begin to adopt alternate strategies for master/slave motion, most commonly, a move-

and-wait strategy is employed to ensure stable motions (Ferrell, 1965). Figure 5–10 illustrates the additional time needed to perform a normalized series of tasks as the delay time increases.

Processing delay arises from the time needed to convert transmitted signals into usable form. Data processing systems spend time encoding and decoding data to avoid noise due to transmission error. The human operator introduces similar delays while processing his or her sensory inputs. The simplest sense-and-move tasks result in significant operator time delays, usually between 113 and 328 msec.

Solutions to the time delay problem measured by the ability to maintain zero-delay proficiency in telemanipulation have proceeded in several directions: (i) task difficulty measurement and reduction, (ii) low-level task autonomy, and (iii) high-level supervisory control. Achieving increased speed in manipulation tasks begins by measuring the difficulty in performing the task relative to the effector employed. These measurements show how to position a task optimally

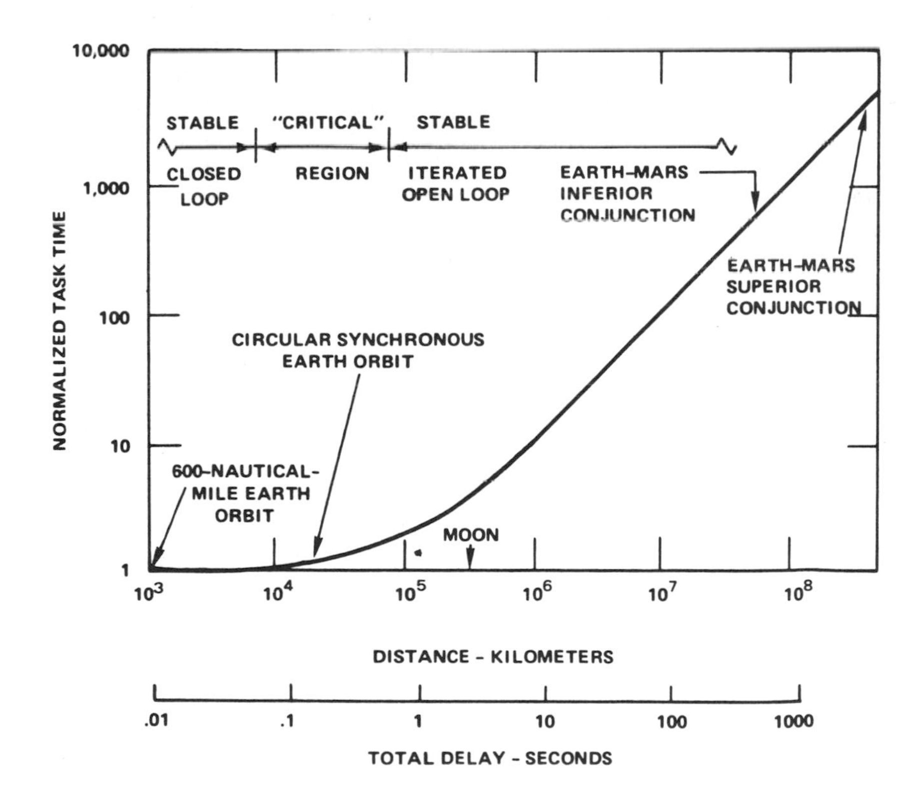


**FIGURE 5–10.** The additional time needed to perform a normalized series of tasks as the delay time increases (Corliss and Johnsen, 1969).

within the workspace of the manipulator (Sturges, 1990a). Analyses of the physics of contact between objects (Ohwovoriole, 1980; Whitney, 1982) and grasping with mechanical fingers (Salisbury and Craig, 1982) have resulted in handling strategies for primitive manipulation tasks. The relatively high-bandwidth tasks of precisely positioning an object can in some cases be controlled locally, avoiding the delays of transmission and processing by the operator. Indeed, the principles of manipulating objects *without* fingers (by pushing and sliding) are becoming known (Mason, 1988; Wang and Mason, 1987). These results have led to the introduction of end effectors and motion plans that guarantee successful task completion in minimum time.

At present the application of these methods to practical teleoperation is limited to the laboratory; however, certain devices such as the remote center compliance (RCC) are available commercially (Lord Corporation, 1989). The RCC simplifies the basic problem of symmetrical peg-in-hole insertion and removal such that no feedback is required for success. Similarly, the "single-freedom-path" strategy of Caine (1985) coupled with a passive mechanism in the wrist accomplishes rapid insertion of nonsymmetrical geometries. We will explore the ranges of tasks that have yielded to such analyses in Section 5.

Research in assembly and disassembly of objects has led to the development of higher-level-motion plans; that is, the paths needed to be followed by a part or an end effector when there is no contact with other things in the environment (Mattikalli et al., 1989). Where models of the objects in the remote location can be built in sufficient detail, supervisory control can be adopted by computing the details of a task and directing the local controller to carry it out. These models are useful, however, only when the relative positions of task and effector remain accurately known. Data integrity suffers when they vary, as will be discussed in the next section.

## 3–3. World Model Shifting

Matching remote locations with local models has been a technique of master/slave teleoperation for decades. Local models that surround the master arm are built to a convenient scale and generally include a large number of features of the remote world. For example, part of the control deck of a remote-service teleoperator system is shown in Figure 5–11. This system performs numerous inspection and maintenance tasks in the hot, radioactive environment of an idle nuclear steam generator. The master arm is seen being manipulated by the operator within a quarter-sphere model of the remote equipment. The remote arm is observed through the monitors, while the smaller-scale master is moved.

In this case, the entire local model is inverted to reduce operator fatigue; the table top upon which the operator is working corresponds to the *ceiling* of the remote site. Key dimensions have been held to (proportional) close tolerances,

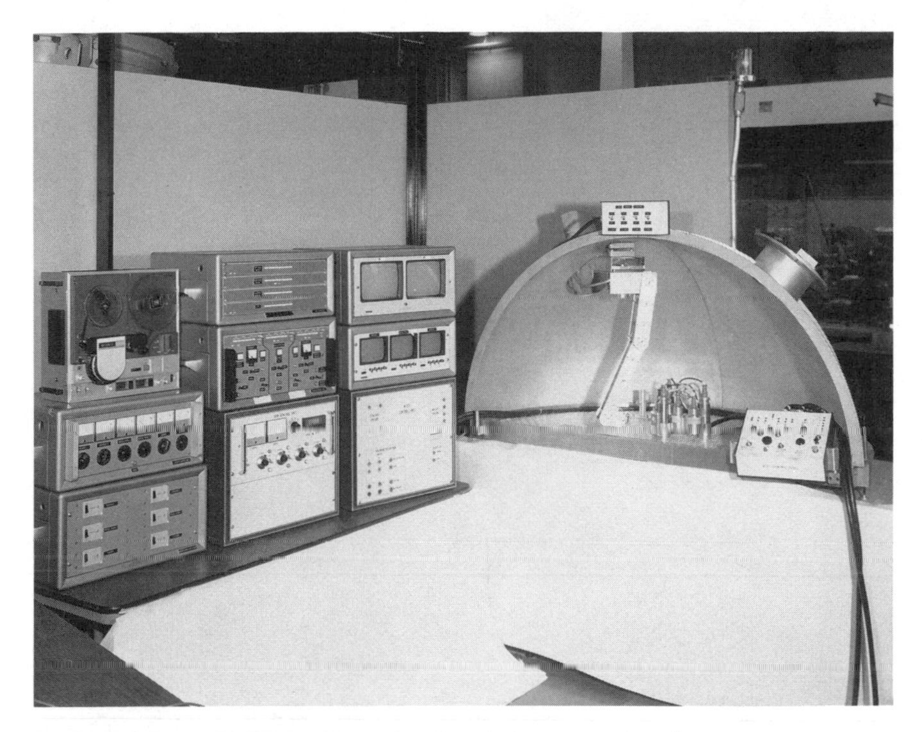

**FIGURE 5–11.** M/STI SAM control deck, a dual remote service teleoperator system.

viz., the features of the steam generator are replicated at the master model to within a few thousandths of an inch accuracy; the slave robot mounting location is remotely measured and adjusted to similar accuracy before maintenance and inspections begin. A miniature "wire-frame" model of the remote end-of-arm tooling attaches to the master arm to guide the operator in positioning tasks. Only the festooning of power cables cannot be replicated, since gravity is essentially reversed! The success of these missions depends on the absence of model shifts between the master and slave. The operator also relies upon frequent references to the monitors, which view the workspace from two nearly orthogonal angles.

System designs that monitor the integrity of the positional relationships among master, slave, and environment are clearly required to ensure personnel and mission safety. In the model-mode system of Figure 5–12, the visual checks of the operator are complemented by several feedback sentinels: (i) tracking error for each servo in the slave (but not the master, since these were not bilateral devices), (ii) position offset error during tool coupling and release, and (iii) binary contact switches to verify motion sequence completion. These built-in checks give the system operators the detailed measurement data they need to detect and correct subtle shifts in the world as they occur.

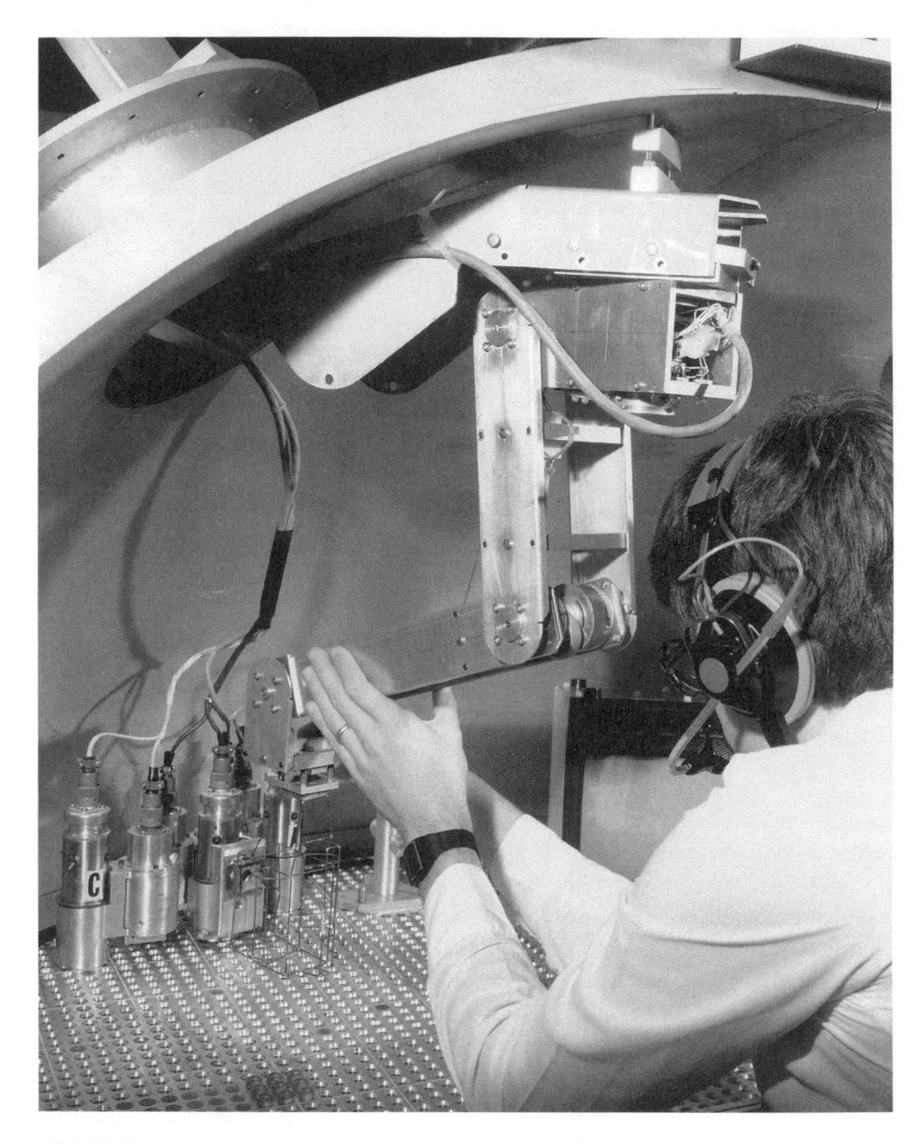


**FIGURE 5–12.** Model mode operation of M/STI and SAM.

## 4. CONTROL DESIGN

Control systems for teleoperators have improved steadily from the analog-controlled bilateral-force-reflecting devices pioneered at Argonne National Laboratory in the 1960s. Stability was guaranteed for simple bilateral manipulators by providing differential velocity feedback with torque feedback to improve "feel" at the master end (Arzbaecher, 1960). More recent models employ similar control. principles, but with digital controllers (Remotec, Inc., 1990). A number of issues regarding the interactions among operator, robot, and environment must be addressed by the control system to enhance mission efficiency, reliability, and safety. The issues discussed here fall under the headings of load factor, transient response, reaction time, and boundary violation and are typically ignored by the industrial robot community. This discussion will indicate the need for multiple modes of teleoperation for enhanced safety and utility as well as the existence of a *mode arbiter,* a process that determines the conditions under which a mode change should take place.

### 4–1. Load Factor

Manipulators and industrial robots share a common problem in the form of control response to a varying payload. The ratio of payload to robot weight is called the *load factor*. The payload of an industrial robot typically measures a few percent of the total active weight or mass of the robot itself, while the space shuttle manipulator is routinely called upon to handle inertial loads many times greater than itself. The load factor of an industrial robot ranges from about 0.01 to 0.10, while the ground-based teleoperator may achieve a load factor near unity. Variations in load for high-load-factor systems create the possibility for a mismatch between the control system manipulator position, irrespective of payload. When the inertial mass center is close to the axis of a revolute joint, the radius of gyration is small. The same robot with arm fully extended exhibits an effective radius of gyration that may be double the "close-in" position. Classical control principles predict that the dynamic behavior of such a mechanical system will vary substantially as the mass components vary (Kuo, 1986). For example, the amount of overshoot in response to a commanded step change in position can vary from zero (critically damped) to 30% of the commanded step when the mass content decreases by a factor of 2.

The above behaviors are mitigated to some extent in bilateral master/slave teleoperators. The master operator feels the inertia of the load and adapts his or her control strategy (e.g., slow down with this load) to maintain stability. Also, his or her musculature acts to extract energy from the remote object when the frequencies are low. For unilateral devices, mission safety depends on control protocols that consider inertias that are comparable to, or even larger than, the slave. Control of the flexible structures of space teleoperators is an area of active research (Book, 1987).

Dynamic response management in the presence of high load factors has been achieved in the laboratory (Kanade and Schmitz, 1985). "Direct-drive" arms (no gearing between actuator and arm) have been constructed that feature very low joint friction and inertia, yet are capable of load factors approaching unity. The control system contains a model of the joint configuration of the arm and a model of the load inertia. It adjusts the gains and damping ratios accordingly and achieves optimum performance everywhere in the workspace. The computational load is high, requiring a multiprocessor system with high reliability and a net throughput of 5 million instructions per second (Mips). Commercialization of this type of control is possible, but not yet realized. This development is mentioned in connection with teleoperators for two reasons: (1) this technology must be applied to future telerobots to realize their full potential and guarantee their safe operation (i.e., avoid instability); and (2) current telerobots are beginning to adopt programmable modes of operation, in which they are expected to perform without the master operator in the loop as a stabilizing agent.

## 4–2. Transient Response

The manipulator as well as its industrial cousin must manage transients in its load. These may arise from internal or external conditions. Examples of external transients include sudden changes in load (e.g., dropping an object) and contact with the environment. Internal transients occur when switching between position control to force control in laboratory robots or when mode switching. The problems of transient response are current areas of active research.

External transients are being addressed by model-reference-adaptive-control (MRAC) methods (Neuman and Stone, 1983). In this approach, a dynamic model of the robot and its load is updated in the controller by measuring shifts in the actual versus expected responses of the system to motion commands. MRAC is also helpful in modeling the changes in inertia that occur as the robot changes joint angles. An explicit model of the robot dimensions and inertia properties is not required *a priori;* the MRAC adapts as needed.

Internal transients caused by changes in mode are less well understood. The modern telerobot offers human-in-the-loop as well as preprogrammed operating modes. The switch from master/slave to programmed control or from position to rate control, independent of signal/command source, gives rise to potentially dangerous transients. Two problems that often occur are unexpected jumps in position at the time of the switch and uncontrollable motions after the time of the switch.

While still an area of research, the following causes of internal transients are identified (Sturges, 1989): (1) mismatch between present state and controller initial conditions, (2) real stored energy in system mechanical/electrical elements, (3) virtual stored energy in controller model of the system, and (4) inappropriate gains or switch-point settings.

The present solution to these transients is simply to bring the robot to a "home" location with no load before every mode switch. Mode switching in linearized system models has been studied by Nett (1988). In this case, the corresponding initial conditions are satisfied as a prerequisite for the mode switching to occur. This approach has yet to be extended to the highly nonlinear, multifreedom telerobotic systems.

## 4–3. Reaction Time

As discussed in Section 3.2, time delays strongly influence the task completion time and the strategy used to carry it out. The advent of multiprocessor control architectures [e.g., NASREM in Lumia et al. (1990)] makes enhanced safety systems possible. Whereas single-mode bilateral telerobots feature an all-purpose "shutdown" mode in which power is removed and brakes are applied, intelligent override could be initiated for error or danger conditions that must occur at a more rapid rate than a human-in-the-loop. For example, the incipient slip of an object in the manipulator grasp is possible with shear force detectors (Howe et al., 1988). Such signals could in principle be displayed to the operator, but it would do little good if the slip actually occurred. The time needed to react to the slip condition would be far too short given the long sensory-motor pathway exisiting in practice. A grasp-mode controller monitoring the sensor and the gripper could be called on to override the master operator force command automatically in the event of small detected slips. Display of override activity would be very valuable to the operator during the planning of a higher-level strategy for handling the object (e.g., place and regrasp).

Similarly, inadvertent contact with a stiff environment could lead to instability by repeatedly bouncing off the wall. A fast digital process monitor could detect such a strike condition through sensors and joint angle errors. An override mode could be switched on to take temporary evasive action or overdamp the main control. Displaying this override would give the operator the extra time needed to react to this condition.

## 4–4. Boundary Violation

Under ideal circumstances, the remote task is planned and carried out, dropping nothing and striking nothing inadvertently. Entry of the robot or its payload into an area already occupied by a hard object is, however, an exceedingly common occurrence for both programmed and human-in-the-loop systems. Indeed, modern teleoperators and industrial robots alike are equipped with shear pins or other mechanical "fuses" to minimize the damage due to the inevitable error. One cannot simply reduce the force capability of the robot to a safe level without removing its ability to perform useful work. Moreover, even a very low force robot can develop significant kinetic energy while making a routine move. As complex and efficient as our human control system is, we still stub our toes and jam our fingers.

Violation of a workspace boundary can occur in two related ways: passing the

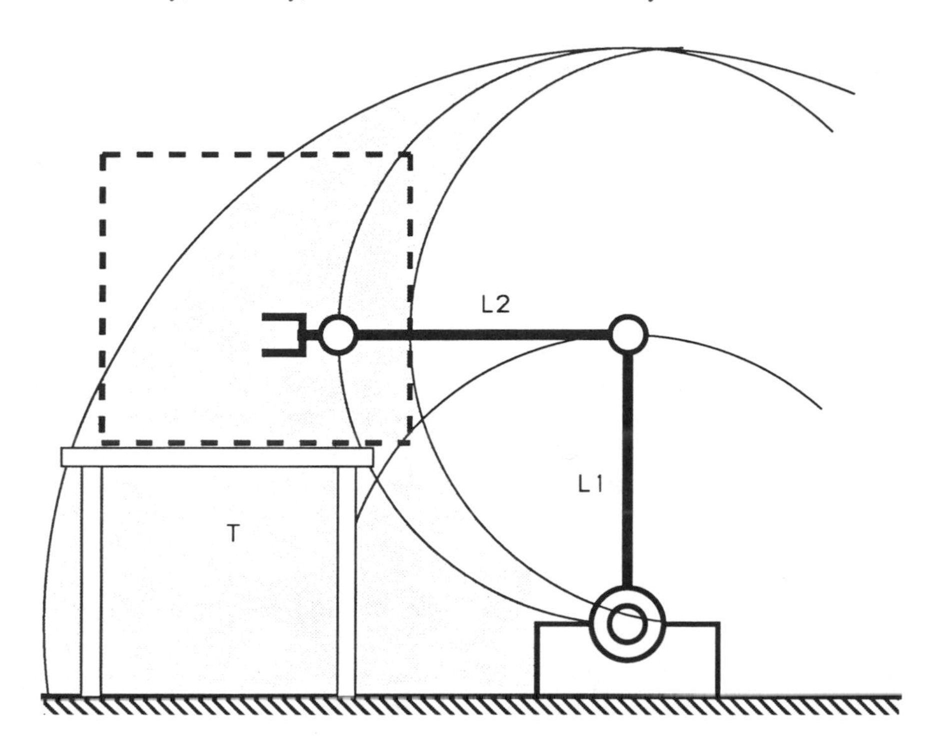


**FIGURE 5–13.** Boundaries in a telerobotic workspace. External boundaries are typically fixed; internal boundaries are more variable.

external limits of the workspace (hitting the walls or invading a neighboring space) or striking an object enclosed within the workspace. The distinctions between the two are not always clear, as shown by Figure 5–13. The workspace T of the slave robot contains a table, which may act as an explicit boundary. The arc defined by the extreme reach of the robot (lengths $L_1$ plus $L_2$) provides an implicit boundary and also imposes limits to the dexterity of the robot near that edge. Attempts at aiding the master operator in obstacle avoidance typically involve finding the robot's position and comparing it to known sets of constraints. This process involves the real-time computations of the robot kinematics and the current "world model," followed by motor control override commands to the slave. Several methods have been tried in the laboratory (e.g., potential gradient method). They depend, in principle, on error-free software and control system operation. Practical controls include axis-by-axis limits, which rarely correspond to the envelope of a real workspace and cannot respond to objects within it.

An alternative approach to external boundary maintenance is given by me-

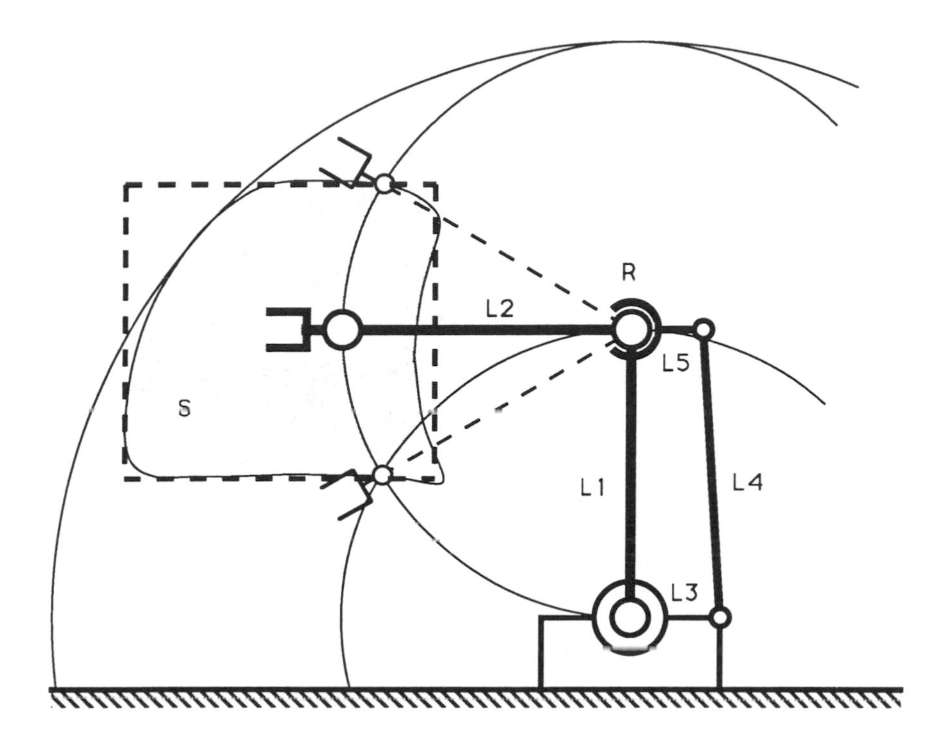


**FIGURE 5–14.** Mechanical computation of workspace boundaries.

chanical computation through mechanisms attached to the robot links and joints as shown in Figure 5–14 (Sturges 1990b). Revolute manipulators feature characteristically spherical extreme reaches. Restricting its reach to a rectilinear workspace (shown within dashed lines) is accomplished by attaching a linkage, formed by $L_3$, $L_4$, and $L_5$, to a set of "hard stops" provided by ring $R$. Since the links subtract a function of the base joint angle, reference to the fixed base is maintained, and the wrist joint becomes bounded, approximately, within the required rectilinear workspace. In general, a smooth external boundary can be defined as a function which constrains a tuple of joint angles. Such functions can be closely approximated mechanically by computing linkages (Hartenberg and Denavit, 1964), which do not depend on control systems for reliable operation. Figure 5–15 shows an example linkage which computes a continuous function of two variables: $\theta_3 = f(\theta_1, \theta_2)$. In the figure, values of $\theta_1$ and $\theta_2$ define the position in the plane of point p, which in turn determines angle $\theta_3$. With all inputs and outputs presumed to be rotations, such a mechanism is readily implemented as an accessory on a revolute teleoperator arm.

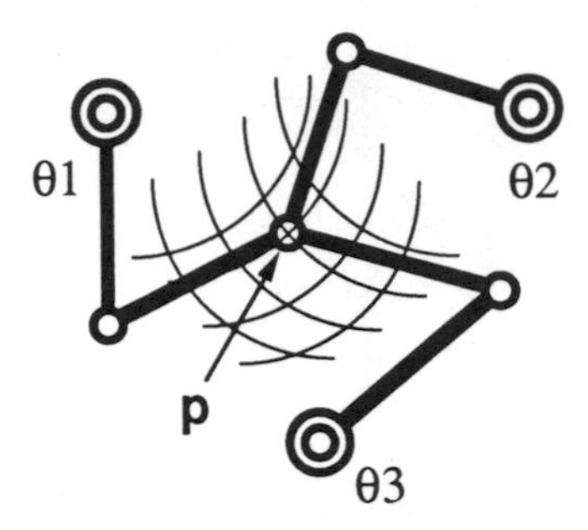


**FIGURE 5–15.** Planar revolute mechanism computing a continuous function of two variables.

## 5. TASK REQUIREMENTS

Perhaps the greatest impediment to teleoperator mission reliability and effectiveness lies with our imprecise knowledge of how to perform the tasks we take for granted in the sensory, "hands-on" world. The primary effort in telerobot development has always been to create a more accurate and precise "telepresence" for the master operator: better vision, kinesthetic sense, and tactile feedback. We need these features because we literally do not know what we are doing when performing a routine manual task. We rely on our manual skills developed over a lifetime of experience. As a telling example of our understanding of routine manual tasks, industrial robots (which are only "remote" in the sense that they are preprogrammed and operate autonomously) have great difficulty in performing dexterous work. Rather, the task is changed to suit their limited abilities. What do we need to know about tasks so that we can direct research in telepresence and begin to work in a remote environment reliably and efficiently? In this section we will examine the kinematics of tasks to determine the qualities of the effector systems that can perform them.

### 5–1. Kinematics and Dexterity

By dexterity we generally refer to skill in the use of the hands, but a skill is a very broad concept and hands usually include in some way all the elements of a manipulative system. Such thoughts associate dexterity with these ideas as follows (Sturges and Wright, 1989):

- A large number of degrees of freedom in the manipulator or hand.
- The ability to achieve fine resolutions with high accuracy.
- The ability to achieve these same resolutions of accuracy at high speed.
- The ability to achieve all of the above in a large working volume so that many parts of the environment can be accessed by one manipulator.

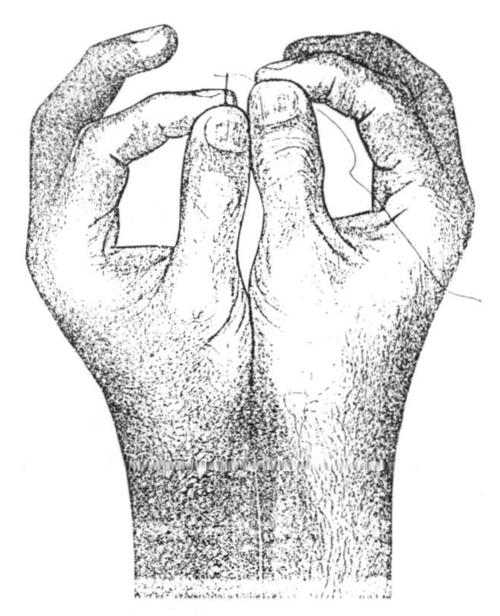

**FIGURE 5–16.** Threading a needle with wrist support.

From a quantitative viewpoint the above observations may be linked to various experiments that have been done in the literature. One simple example (Fitts, 1954) consists of moving a point in one dimension between two target locations in width $w$ and separation $s$ (Figure 5–16). Fitts has defined and measured the index of difficulty (ID) for this test as:

$$ID = \log_2( s / w ). \quad (5\text{–}1)$$

Measurements over large numbers of human subjects show that the task time is linearly related to this index of difficulty by a factor of about 100 msec/bit (Figure 5–17). A mechanism capable of positioning to an equal or better combined resolution will thus be capable of performing the task, if the appropriate scale is considered. Therefore, an index of difficulty for a task is analogous to an index of capability for an effector. The two are equal when the effector is just sufficient to accomplish the task. We will use the term *index of difficulty* (ID) for both the task and the effector.

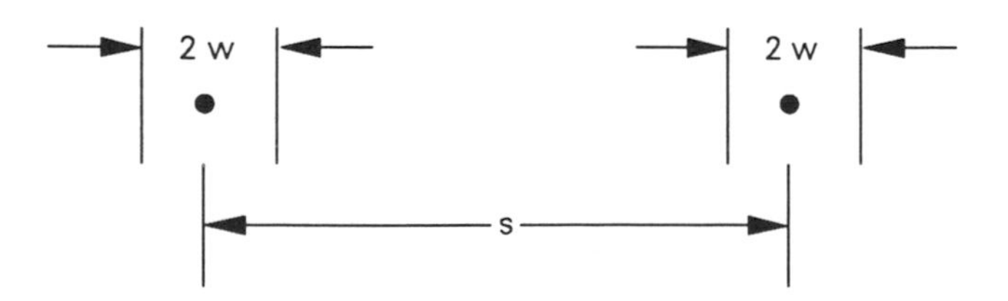


**FIGURE 5–17.** Schematic of Fitts's tapping task: subject makes a jot between each set of bars alternately at maximum speed and minimum error.

Expression (5–1) tells us something about the difficulty of performing a positioning task, but does not address forces, or whether the directions along which a telerobot features a given resolution are appropriate to the task. Active control of compliance (Whitney, 1987) relates forces and displacements, but compliance alone does not guarantee dexterity. We must also quantify the force resolution about each axis. This can be expressed as a nondimensional measure of sensitivity:

$$ID = \sum_i \log_2 (1/f_i), \qquad (5\text{–}2)$$

where $f_i$ is the minimum applied force along an independent axis of an effector for which the unit force is a maximum. When we apply this measure to the tasks of threading a needle, closing a bureau drawer, or setting a utility pole, each task might involve an equal measure of dexterity by a force (or displacement) measure, and yet the actual levels of force and distance are very different.

A measure of dexterity based on the foregoing effector kinematics and task physics (Sturges and Wright, 1989) will now be applied to a fundamental manipulation task. If we take into account the geometric constraints of our effector (e.g., extreme reach), we can determine if a given telerobot can accomplish the task by comparing the corresponding ID along each axis $i$. A summation of ID values will not work, since each axis must compare favorably for guaranteed success. For specific applications of a given effector to a given task, we can determine the *next dexterity* of the task/effector system from the minimum of the differences between the capability of the effector along each axis (its measured indices of difficulty) and the difficulty of the task along the same axes (its required indices of difficulty), viz.,

$$\text{Dexterity}_{net} = \underset{i}{\text{MIN}} \, (\text{ID}_{i \text{ effector}} - \text{ID}_{i \text{ task}}). \qquad (5\text{–}3)$$

Note that dexterity is maximized for high effector capability and minimized for high task difficulty (and conversely). Positive values of dexterity indicate successful completion of the task; negative values of dexterity suggest that the effector is not fully capable of the given task.

## 5–2. Dexterous Manipulation

To illustrate the procedure of finding the net dexterity of a task/effector system, let us examine the two-dimensional peg-in-hole problem. The geometry of this manipulation task is shown in Figure 5–18 wherein a plane peg is elastically supported. It has been shown (Whitney, 1982) that at least two conditions on contacts must be considered to describe this task and guarantee success, viz., a geometric condition known as wedging (when a bureau drawer gets stuck) and a

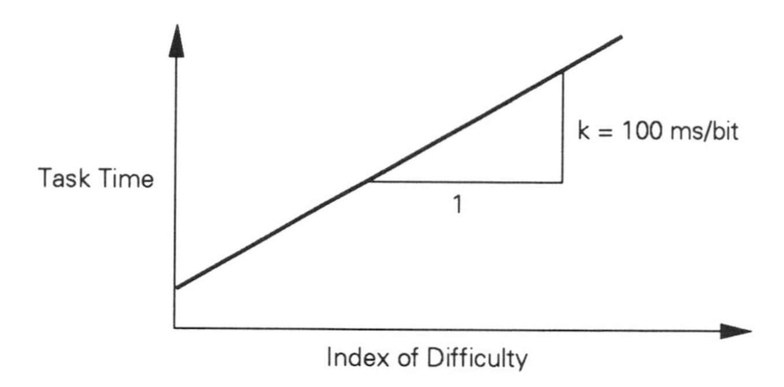


**FIGURE 5–18.** Results of Fitts's experiments: task time is proportional to the index of difficulty. Human subjects fall into a narrow range of performance around 100 msec/bit.

force application condition called jamming (when chalk screeches on a blackboard).

To avoid wedging, the lateral error $\epsilon_X$ of the effector must be less than the chamfer width $W$. Furthermore, the nature of the compliance supporting the peg leads to a requirement on the initial angular error $\theta_0$. Drake et al. (1977) show the relationship between these values, the clearance $c$ between the peg and hole, and the friction $\mu$ at their surfaces. These two errors can be combined into a single angular error bound $d\theta_t$ within which the effector must operate:

$$d\theta_t = \frac{c/\mu}{kS + 3}, \tag{5–4}$$

or

$$d\theta_t = W/k, \tag{5–5}$$

whichever is smaller. (The value $S$ combines the ratio of angular and lateral compliances with the insertion distance.) Length $k$ is derived from the geometry of the manipulator and its joint angles. The joint angle error $d\theta_e$ is simply related to the starting errors by the manipulator Jacobian. The relationship gives

$$\theta_0 = 3d\theta_e, \tag{5–6}$$

$$\epsilon_x = kd\theta_e, \tag{5–7}$$

where k $= -L\,[\sin\theta_1 + 2\sin(\theta_1 + \theta_2)]$ and $L$ is the length of the manipulator link as shown in Figure 5–19. For the position of the arm shown in the figure, $\theta_1 = \pi/4$ and $\theta_2 = -\pi/2$. At this point the value $k = 0.707L$. In general, the value of $k$ will vary with manipulator position and orientation, thus directly affecting the dexterity of the effector with respect to the given task. When the index of

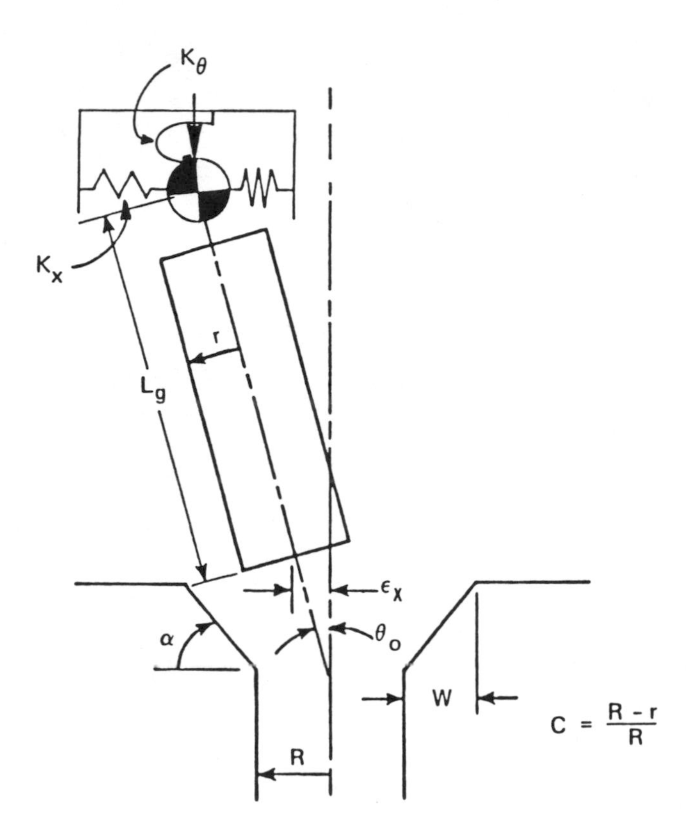


**FIGURE 5–19.** Definition of terms for geometric analysis of wedging. [After Whitney (1982).]

difficulty for angular error is referred to $2\pi$ radians, the dexterity of the system (in bits) can then be found from Eq. (5–3):

$$\text{Dexterity}_{\text{net}} = \log_2 2\pi/d\theta_e - \log_2 2\pi/d\theta_t = \log_2(d\theta_t/d\theta_e). \quad (5\text{–}8)$$

A value of effector angular error $d\theta_e$ that exceeds the allowable task angular error $d\theta_t$ will indicate the possibility of failure. The net dexterity will then be negative.

The foregoing suggests a means to examine dexterity as a function of the workspace. Consider the workspace of the effector of Figure 5–20, and the peg-in-hole task positioned variably within it. A map of dexterity "potentials" (as shown in Figure 5–21), if found by solving for the joint angles and applying Eqs. (5–4) to (5–8). Higher values show a margin of dexterity; these are the "sweet spots" for wedging avoidance in this task/effector system. It is intuitively

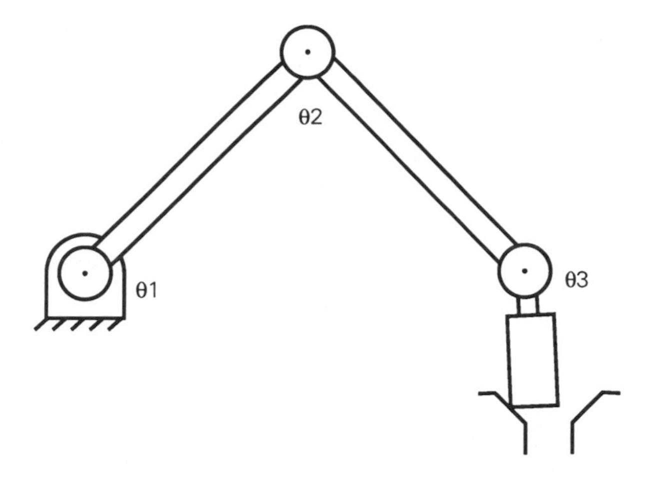


**FIGURE 5–20.** Articulated effector and peg-in-hole task geometry.

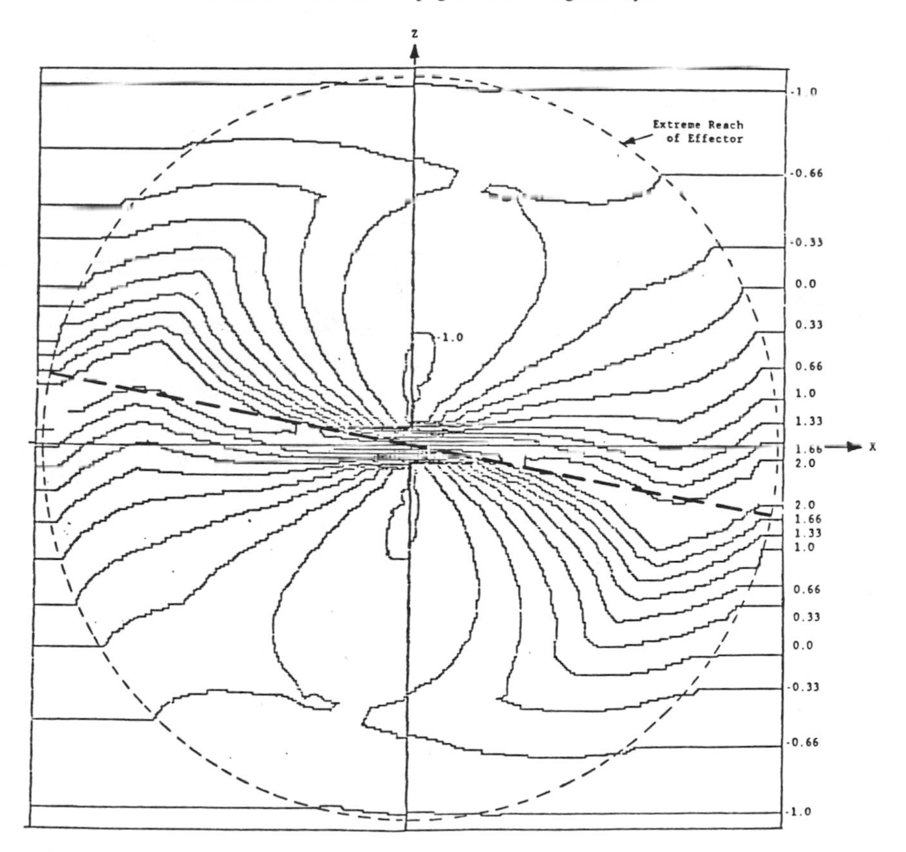


**FIGURE 5–21.** Contours of equal dexterity in world coordinates for the articulated effector and task of Figure 5–20.

satisfying that these areas occur near the first joint of the arm. The dexterous workspace is bounded by contours of zero dexterity: those locations beyond which the task cannot be performed reliably.

## 6. SUMMARY

As the teleoperator evolved in strength and sophistication, the issues of reliability and safety became part of the success of its mission and its design. Bilateral force reflection put the operator in the loop in ways which could cause injury in the event of a system malfunction. The many possible failure modes of a teleoperator system fall into four principal domains: system integrity, data integrity, control design, and task requirements.

The system integrity of a teleoperator can be affected by loss or reductions of functions. A loss of function condition may occur due to structural, power, or signal failures. Such a loss of function may not lead to complete system or mission failure if devices designed for detection, redundancy, and override are incorporated. The more subtle reductions of function caused by stress, strain, and kinematic drift can be compensated in most cases through autocalibration and path-planning techniques.

The data integrity of a teleoperator control system may be lost due to device degradation, time delays in a perfect system, or shifts in the workspace model that go undetected. Impedance detection has been successfully employed for errant feedback and control devices. System knowledge in the form of mode control and sensing experts have been shown to ameliorate the effects of delays and model shifts.

The success of a teleoperator system depends critically on the allocation of parameters in its control system. Changes in inertial or gravitational loads are effectively handled by model-reference adaptive control methods. Reactions to transients are problematic for many control systems, especially when shifting control modes, unless the control system design includes initial condition tracking. The reaction time of an operator may not be sufficiently short for all expected system behaviors. Sensing of such conditions and the ability to shift from manual to reflex modes through a mode arbiter could enhance the apparent abilities of the operator. The increasing speed of digital control has made world coordinate-based boundary maintenance feasible. Mechanical kinematic computation provides the passive "back-up" in case of system, actuator, or feedback element failure.

Mission reliability and task speed may be enhanced when the kinematic and dynamic elements of the tasks themselves are modeled with the particular dimensions of the effector supplied to execute them. Dexterity analysis provides a means to assess the efficacy of a mission by evaluating the net dexterity of a system within its workspace under varying conditions.

In each of the above problem domains, the operator is presumed to be a constant and predictable entity. As we push the frontiers in telemanipulation, we will find that this approximation does not hold very well. The performance of human-in-the-loop is one of the greatest sources of mission variation, and yet he or she remains indispensable in all but the simplest structured environments. We can look forward to reducing this variability and enhancing the safety of telerobotics as we expand, develop, and employ the techniques outlined in this chapter.

**References**

Arzebaecher, Robert C. 1960. "Servomechanisms with Force Feedback," Argonne National Laboratories, ANL-6157.

Blackburn, J. F., Reethof, and G., Shearer, J. L. eds. 1960. *Fluid Power Control*. Cambridge, MA: MIT Press.

Book, W. 1987. Robust scheme for direct adaptive control of flexible arms, *Winter Annual Meeting of the ASME Dynamic Systems and Control Division,* Boston, MA.

Caine, Michael E. 1985. Chamferless assembly of rectangular parts in two and three dimensions, Master of Science Thesis, MIT.

Cooling, J. E. 1986. *Real Time Interfacing*. United Kingdom: Van Nostrand Reinhold, Ltd.

Corliss, W. F., and Johnsen, E. 1967. *Teleoperators and Human Augmentation*. AEC-NASA Technical Survey, Washington, DC: NASA Office of Technology.

Corliss, W. R., and Johnsen, E. 1968. *Teleoperator Controls*. AEC NASA Technical Survey, Washington, DC: NASA Office of Technology.

Drake, S. H., Watson, P. C., and Simunovic, S. H. 1977. High speed assembly using compliance instead of sensory feedback, *Proceeding of the 7th International Symposium of Industrial Robots,* Tokyo, Japan, pp. 87–97.

Ferrell, W. R. 1965. Remote manipulation with transmission delay, NASA TN D-2665.

Fitts, P. M. 1954. The information capacity of the human motor system in controlling the amplitude of movement. *J Experimental Psych* 47(6).

Goertz, Ray C. 1949. "Master-Slave Manipulator," Argonne National Laboratories, AEC ANL-4311.

Goertz, Ray, C. 1954. "Mechanical Master-Slave Manipulator," *Nucleonics* 12(Nov.):45–46.

Greenaway, P. M., and Sturges, R. H. 1979. HEF manipulator systems reliability testing programs and increased capacity investigation. Westinghouse Advanced Reactors Division Report DDI-79-1018.

Hartenberg, R. S. and Denavit, J. 1964. *Kinematic Synthesis of Linkages*. New York: McGraw Hill.

Howe, R. D., Kao, I., and Cutkosky, M. R. 1988. The sliding of robot fingers under combined torsion and shear loading. *IEEE 1988 Conference on Robotics and Automation,* Philadelphia, PA.

Jakuba, Stan R. 1986. Failure mode and effect analysis: a tool for reliability planning and risk evaluation. *Fall National Design Engineering Show and Conference,* New York, NY.

Kanade, T. and Schmitz, D. 1985. Development of CMU direct-drive arm II. *Proceedings of the American Control Conference*.

Kanade, T., Mitchell, T., and Whittaker, W. L. 1989. 1988 year end report: autonomous planetary rover at Carnegie Mellon University. Carnegie Mellon University Robotics Institute TR-89-3.

Khosla, P. K. 1989. A system to determine assembly sequences. Technical Report, Carnegie Mellon University Engineering Design and Research Center.

Kuo, B. 1986. *Automatic Control Systems*. Englewood Cliffs, NJ: Prentice Hall.

Lee, S.-Y. and Shearer, J. L. 1955. Development of a miniature electrohydraulic actuator, *Trans ASME* 77:1077–86.

Lord Corporation 1989. Remote compliance center devices. Industrial Automation Division, Suite 300, 118 MacKenan Drive, P.O. Box 8200, Cary, NC 27512-8200, (919) 469–2500.

Mason, M. T. 1988. How to push a block along a wall, *NASA Conference on Space Telerobotics*.

Mattikalli, R. S., Khosla, P. K., and Xu, Y. 1989. Subassembly identification and motion generation for assembly: a geometric approach, M.S. Thesis, Carnegie Mellon University Engineering Design and Research Center.

Menq, C. 1988. The dextrous workspace of simple manipulators. *IEEE Robotics and Automation* 4(1):99–102.

Mosher, R. S. 1967. Handyman to Hardiman. SAE Paper 670088.

Lumia, R., Fiala, J., and Wavering, A. 1990. "The NASREM robot control system and testbed. *International Journal of Robotics and Automation* 5(1).

Nett, C. N. 1988. An explicit formula and an optimal weight for the two-block structured singular value interaction measure. *Automatica* 24:261–265.

Neuman, C. P. and Stone, H. 1983. MRAC control of robot manipulators. *Proc. 3rd Yale Workshop on Applications of Adaptive System Theory,* K. S. Navendra, Ed., Yale University, New Haven CT, 1983, pp. 203–210.

Ohwovoriole, M. S. 1980. An extension of screw theory and its application to the automation of industrial assemblies. Stanford Artificial Intelligence Laboratory Memo AIM-338.

Paul, R. P. 1981. *Robot Manipulators*. Cambridge, MA: MIT Press.

Paynter, H. M. 1953. Electrical analogies and electronic computers: surge and waterhammer problems. *Transactions of the American Society of Civil Engineers* (118): 962–1009.

Railbert, M. H., Brown, H. B., Chepponis, M., Hastings, E., Koechling, J., Murphy, K. N., Murphy, S. S., and Stente, A. J. 1983. Dynamically stable legged locomotion. Carnegie Mellon University, Robotics Institute, RI-TR-83-20.

Remotec, Inc. 1990. RM-10A product description, 114 Union Valley Road, Oak Ridge, TN 37830, (615) 483–0228.

Sahar, G. and Hollerbach, J. M. 1986. Planning of minimun-time trajectory for robot arms. *International Journal for Robotic Research* 5(3):90–100.

Salisbury, J. K. and Craig, J. J. 1982. Articulated hands: force control and kinematic issues, *International Journal of Robotics Research* 1(1):4–17.

Song, S. M. and Waldron, K. J. 1988. *Machines That Walk: The Adaptive Suspension Vehicle*. Cambridge, MA: MIT Press.

Stone, H. W. 1986. Kinematic modeling, identification, and control of robotic manipulators. Ph.D. Thesis, Electrical and Computer Engineering Department, Carnegie Mellon University.

Sturges, R. H. 1989. MFP Control document, Westinghouse Productivity and Quality Center.

Sturges, R. H. 1990a. A quantification of machine dexterity applied to an assembly task. *International Journal of Robotics Research* 9(3).

Sturges, R. H. 1990b. Mechanical computation of robot workspace envelopes. Carnegie Mellon University Robotics Institute TR-90-57.

Sturges, R. H. and Wright, P. K. 1989. A quantification of dexterity. *Journal on Robotics and Computer Integrated Manufacturing* 6(1):3–14.

Tesar, D. 1989. Thirty-year forecast: the concept of a fifth generation of robotics—the super robot. *Manufacturing Review*, (1):16–25.

Wang, Y. and Mason, M. T. 1987. Modelling impact dynamics for robotic operations. *Proceedings, Fifth International Symposium on Industrial Robotics and Automation*, pp. 678–685.

Whitney, D. E. 1982. Quasi-static assembly of compliantly supported rigid parts. *Journal of Dynamic Systems, Measurement and Control* 104(3):65–77.

Whitney, D. E. 1987. Historical perspective and state of the art in robot force control. *International Journal of Robotics Research* 6(1).

Wright, P. K. 1985. A manufacturing hand. *Journal of Robotics and Computer Integrated Manufacturing* 2(1):13–23.

# 6

# A Multilevel Robot Safety and Collision Avoidance System

James H. Graham, Porter E. Smith, and Ramanchandra Kannan

## 1. INTRODUCTION

The closely related topics of robot safety and robot collision avoidance in dynamic, nondeterministic environments comprise a significant, and possibly even a crucial, challenge to robotics researchers. Many initial applications of robots in industrial settings have been in very simple repetitive tasks, such as material transfer, spot welding, and spray painting, where the robot manipulator is typically anchored to a stationary base, its sequence of movements is reasonably predictable, and its reach is limited. For these applications, standard industrial practices, such as warning signs, barriers, and interlock devices, may be employed to keep untrained personnel away from an operational robot.

There are however several imminent developments in the field of robotics that will upset this situation of relative safety, as robots begin to be used in applications involving more direct contact with humans. As these new trends become more completely realized, it seems obvious that operator training and restriction of access will no longer suffice to maintain adequate conditions of safety, and more active forms on robot safeguarding will have to be developed (Sugimoto and Kawaguchi, 1983). This chapter presents a multilevel system for improving robot safety by the use of active sensory data from the robot's sensory system and includes provisions for preprocessing of the data, integration of data from disparate sensors, and high-level decision making.

The subject of collision avoidance in deterministic environments has been extensively studied and reported because of its key importance in path planning for automatic off-line programming of robot tasks. However, there has been a relatively small amount of research on the problems of robot safety and dynamic collision avoidance. Research groups at National Institute of Standards and Technology (formerly the National Bureau of Standards) (Kilmer, 1982) and Rensselaer Polytechnic Institute (Meagher et al., 1983; Graham et al., 1986) have done preliminary investigations of the use of presence-sensing devices for robot safety. NIST researchers (Kilmer et al., 1984) have also investigated the use of a redundant set of encoders and a separate watchdog safety computer to enforce restricted zones of the workspace. Bennekers and Ramirez (1986) have investigated the use of computer vision techniques. Researchers at West Virginia University and the National Institute of Occupational Safety and Health (Sneckenberger et al., 1987) have investigated a combination of light curtains, pressure mats, and ultrasonics. Sugimoto and Kawaguchi (1983) have used fault-tree-analysis techniques to assess robot hazards.

This chapter is primarily concerned with how robot safety can be improved by the use of active sensory data. There are two principal components to achieving this goal: (1) the processing and reduction of the raw sensory data to extract key features of the workspace and (2) the intelligent use of these extracted features in making decisions about safe operation. Section 2 provides an analysis of the functional requirements for robot safety. Section 3 discusses the characteristics of typical sensory devices used for safeguarding. Section 4 discusses an approach to the sensory processing problem using belief functions and the Dempster–Shafer theory of evidence. Section 5 presents an approach to the intelligent decision making based on an object-oriented representation of the safety system.

## 2. FUNCTIONAL REQUIREMENTS FOR A SAFETY SYSTEM

A successful robot safety system must satisfy a stringent list of both technical and economic requirements. The primary technical requirement is that it should provide reliable detection of intruders and obstacles, while being highly immune to false alarms. It should be rugged enough to survive an industrial environment and should provide fail-safe operation (i.e., the safety system should fail in a manner that the robot is disabled). It should be inexpensive relative to the cost of the robot and should be easy to install and operate. While this chapter is primarily concerned with meeting the primary functional objective of accurate detection, it is important to keep in mind the other requirements.

In developing a robot safety system, it is necessary to consider the regions over which safety should be provided and to specify the appropriate responses within these regions. A good starting point is the three safety regions identified by the NIST (Kilmer, 1982) as follows:

Level I—workspace perimeter protection.
Level II—protection within the workspace.
Level III—protection very near the robot.

This is a reasonable hierarchy of protection for a stationary industrial robot, reflecting the increase in hazard with proximity to the robot.

Level I safety would typically be achieved in an industrial setting by a physical barrier, such as a woven wire fence, or by a perimeter-detection device such as a photoelectrically sensed light curtain.

Level II is defined to be the reachable workspace volume of the robot excluding a small volume immediately surrounding the robot itself. Within Level II an intruder is within reach of the robot, but not in imminent danger of being struck.

Level III is defined to be the volume immediately around the robot. In a simple model, this might be a fixed distance, such as 6 in. In a more complex model, this region would vary with the velocity of the robot.

No formal standards have been established for the response to violations of the various safety levels. A reasonable set of responses would be a visual or audible alarm, with possible limitation on speed of the robot, when anyone enters the volume protected at Level II. Violations at Level III require an immediate emergency stop.

In a hierarchically organized safety system, each sensor is connected to a coordinating unit for its particular sensor group. Typical sensor groups would comprise ultrasound ranging devices, passive and active infrared sensors, capacitive and pressure presence sensing, and other units as determined by the specific application. The coordinator would direct the operational performance of the sensors (such things as firing order, reset times, etc.) and would receive and preprocess the sensed information from these units.

Certain sensory groups provide information that is vital to detection of safety information at the third safety level. (A good example is the use of a capacitive antenna at or near the robot wrist.) In these cases, the local coordinator must monitor the sensed information for violations and very quickly determine if an emergency response is needed. This emergency shutdown request must be directly transferred to the robot control system, and even then the robot is likely to move several inches before motion is halted.

By contrast, other sensory groups provide information well in advance of any direct robot hazard. (As examples, consider photoelectric barriers around the periphery of the robot, passive infrared perimeter detectors, and long-range ultrasound ranging detectors.) These sensors can be used in the emergency shutdown detection mode, but this can lead to many situations in which the robot is stopped when no real hazard has developed. A two level approach for dealing

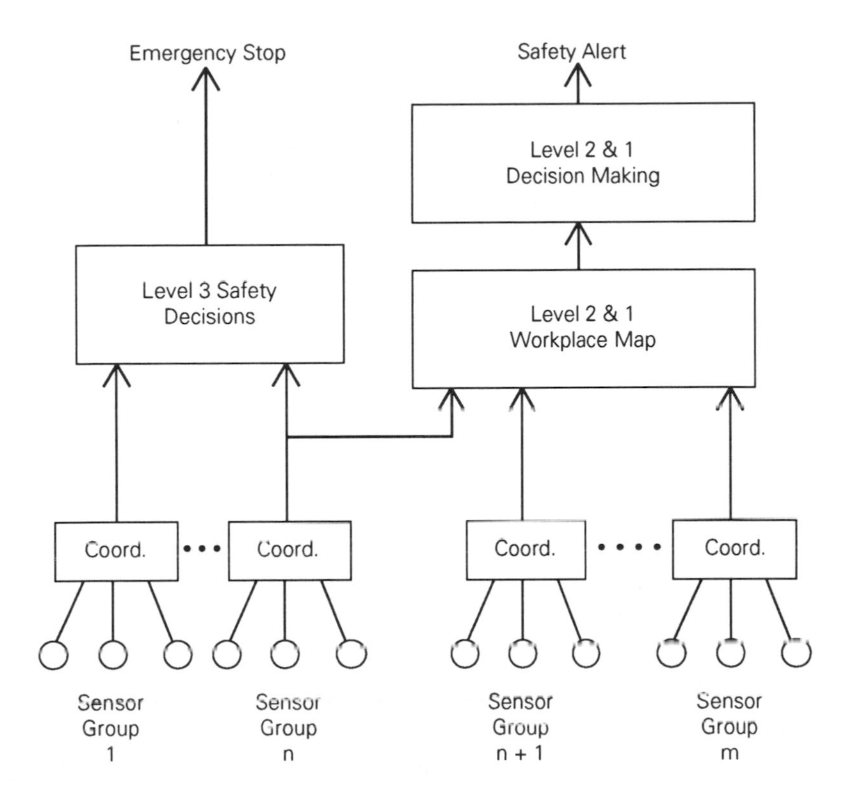


**FIGURE 6–1.** Hierarchical safety system control structure.

with this type of situation would be: (1) local coordinator information is combined with other information to form and update a map of the workstation and (2) decision making about safety alerts or stops is made by the safety decision maker based on information from this map. This approach is shown in block form in Figure 6–1.

## 3. SENSORY SYSTEMS FOR SAFEGUARDING

This section will provide an overview of some candidate technologies for safety sensing instrumentation of industrial robots. Selection of mounting positions is a significant aspect of all sensor applications for robot safety, and varies with the type and model of robot, and the application task. In some cases it is possible to mount the sensor in such a way that no fixed obstacles come within the sensor's range, while in other cases this is not possible.

## 3–1. Ultrasound Sensors

Ultrasound sensing is based on producing a high-frequency (above 20 kHz) sound wave, transmitting this sound wave toward a target, and measuring the time interval until a reflection is detected at the source. The distance to the reflecting target is linearly related to the observed time delay by the speed of sound in air. Ultrasound sensing is used in commercial intrusion detectors, for focus control in instant cameras, and for industrial ranging and gauging. Ultrasound transducers are relatively lightweight and inexpensive, and sophisticated integrated electronic control units have been developed for them. Transducers with detection ranges of a few inches to 30 ft or more are commercially available.

## 3–2. Capacitive Sensors

Capacitive sensors are currently used in the factory environment as intrusion-detection devices around presses and stamping machines to protect workers from inadvertent injury. The sensing unit consists of a radio-frequency generator, a control unit, a coupler, and an antenna. An electromagnetic field is produced in the antenna, which acts as one element in a balanced bridge circuit. The control unit is able to detect small changes in the capacitance between the antenna and ground from this bridge circuit. When an intruder approaches the antenna, the change in dielectic constant changes the capacitance and triggers an alarm circuit. Typical detection ranges are about 18 in.

## 3–3. Infrared Sensors

Humans emit infrared radiation in a well-defined spectral range of about 8–14 μm, and many commercial infrared detectors have been developed that are tuned to these wavelengths. In the pyroelectric detector model, a small ferroelectric crystal is enclosed in a metal container having a small window. Any increase in the radiant energy entering through the window results in the heating of the crystal, and produces an electric current. Most commercial infrared detectors are equipped with sophisticated optical systems that focus different fields of view into the detector window, resulting in a set of detection zones, referred to as "fingers." Typical units have 12 finger pairs, each about 2° in width, which cover a range up to 40 ft.

## 3–4. Microwave Sensing

Velocity-sensing microwave units operate on the principle that when a wave is reflected from a moving object, the frequency of the reflected wave is increased (or decreased) by an amount proportional to the object's speed. These units are used commercially for intrusion detection and to control automatic doors.

It was determined that the microwave lobe pattern is large enough to cover the workspace of a large industrial robot with a single unit. Owing to this good coverage, and the weight of the unit, it was mounted off the robot. This unit complements the other sensors by providing velocity information about objects in the workspace. It does tend to be somewhat sensitive to vibrations.

### 3–5. Other Sensory Systems

Several other sensor systems have been investigated for use in robot safety applications, including pressure-sensitive floor mats, photoelectric light curtains, and computer vision. Floor mats and light curtains are primarily for perimeter-penetration detection and are largely effective at this task. While some preliminary investigation has been performed on the use of computer vision for robot safety, this mode has not been widely used because of cost and the general slow speed of image-processing algorithms.

## 4. INTEGRATION OF SENSORY DATA

As discussed in the previous section, fusion of sensory data at the intermediate processing level, while critical to many advanced automation applications, has received relatively little research attention to date. By the hierarchical structuring principle, this level should exhibit intermediate precision and intelligence. It should serve as a condensing function for higher-level modules by consolidating and removing unneeded numerical detail. However, it should maintain a rigorous mathematical basis to enforce completeness and consistency of the data processed. There are many approaches that might be taken to comply with these conflicting requirements (Albus, 1981; Saridis, 1979; Shafer, 1976; Zadeh, 1978). The purpose of this section is to present one mathematical formulation for this fusion process.

Formally, let $S = \{ s_1, s_3, \ldots, s_n \}$ be a finite set of sensors that are providing information about conditions within the working space of a robot. Let $v_i(k)$ be the value returned by sensor $s_i$ at a particular sampling instant. While $v_i$ may in general be either a continuous or discrete variable, it will be assumed in this formulation that $v_i$ is discrete and comes from a finite set of values. Let $V_i$ be the set of possible values from sensor $i$ and let $V = V_1 \cup \cdots \cup V_n$ be the set of all possible sensor values.

Let the set of attributes about which information is being perceived be enumerated in a sensory configuration space denoted by $G = \{g_1, g_2, \ldots, g_m\}$. The attributes measured by sensors may be global properties of the robot workspace, such as ambient temperature, ambient illumination levels, etc., or they may be localized properties, such as the force exerted at the end effector or the presence (or absence) of an object at a specified point in the workspace.

Let $\gamma$ be the function that relates the measured sensor value and the condition of the sensory configuration space:

$$\gamma : V \times S \times G \rightarrow \mathrm{H}. \tag{6–1}$$

Thus a particular value of $H$, say $h_{i,j,k}$, represents the state of the $k^{\text{th}}$ sensory attribute as reflected by the $j^{\text{th}}$ input value of the $i^{\text{th}}$ sensor. This function provides the mechanism by which the raw sensory value begins to be interpreted into useful information. The function may be as simple as a linear conversion factor or may be a complicated nonlinear operator. In the latter case, it may be desirable to store the function as a lookup table.

If the lookup table approach is used, Eq. (6–1) implies a very large amount of computer storage. In practice, many of the table entries will not be of interest, which implies the use of sparse-data storage techniques, involving several small tables. An indication of the coupling between sensory systems can be given by the following sensory coupling function:

$$\kappa : S \times G \rightarrow I, \tag{6–2}$$

where $I_{ij}$ is 1 if sensor $i$ provides information about attribute $j$, and is 0 otherwise. If the matrix implied by the $\kappa$ function is an identity matrix, then the sensors are measuring unrelated characteristics, and there is no need for sensory fusion.

The goal of the intermediate sensor fusion level is to combine the $H$ maps generated by the individual sensing units into some sort of decision space $D$ as indicated by the following:

$$\xi : H_1 \times H_2 \times \cdots \times H_n \rightarrow D. \tag{6–3}$$

There are several mathematical approaches for dealing with uncertainties when they are encountered in the $\xi$ mapping described above. The author has chosen the approach of mathematical evidence as described by Shafer (1976) for initial investigation. In particular, this approach has the advantage of assigning confidence or belief values to subsets of the sensor configuration set $G$. As will be explained in the case study section, this is a mode that is closely related to the information provided by several types of sensors. The combination of belief functions derived from different sensors is done by Dempster's rule of combination (Dempster, 1967).

Briefly stated, the Dempster–Shafer theory deals with two types of belief; a basic belief (or probability) assignment $m$ defined on all subsets of the frame of discernment (sample space) $\Theta$ by

$$m : 2^{\Theta} \rightarrow [0,1] \tag{6–4}$$

subject to the constraints

$$m(\phi) = 0 \quad \text{and} \quad \sum_{A \subset \Theta} m(A) = 1 \tag{6–5}$$

and derived (or total) belief functions for any event

$$\text{Bel}(A) = \sum_{B \subset A} m(B). \tag{6–6}$$

An advantage of belief functions is that two belief functions $m_1$ and $m_2$ defined over the same frame of discernment can be combined by Dempster's rule of direct combination as follows:

$$m(A) = \frac{\sum_{A_i \cap B_i = A} m_1(A_i) m_2(B_j)}{1 - \sum_{A_i \cap B_j = \phi} m_1(A_i) m_2(B_j)}. \tag{6–7}$$

The numerator is the sum of all subset contributions to $A$ and the denominator acts as a normalization factor by summing all components from empty intersections.

This evidence-based approach to sensory fusion has a number of qualitative and quantitative advantages as a tool for the coordination of multiple-sensory systems. It has a mathematical rigor that allows it to be used readily with probabilistic estimates of sensory performance. It has a consistent rule for combination that is fully associative and commutative, so that the order of combination is irrelevant. It allows the inclusion of nonconclusive or ambiguous evidence (ignorance) without biasing the results. This approach seems to match well with physical intuition about sensory processing.

The major disadvantage of the evidential approach is the potentially large amount of computation required. The power set assignment of basic beliefs implies a worst case exponential explosion in computation. Fortunately many practical problems involve only small subsets of the power set, and computational problems are avoided with a careful selection of algorithms and data structures.

In the special case of proximity sensing the generalized models for sensory fusion given by Eq. (6–1), (6–2), and (6–3) can be specialized. The attribute space $G$ becomes a set of physical points in the robot workspace. The coupling function indicates the locus of points over which a particular sensor provides information. The $\gamma$ function provides either a binary indicator of the condition of workspace points (free or occupied) or, in the more sophisticated case, an indication of the confidence, probability, or belief of occupancy.

As simple examples of sensory fusion in this application, consider the two cases of detection of an object in the robot workspace by area-type sensors as shown in Figure 6–2. Clearly if both sensors indicate a reasonable degree of certainty of detection, as indicated by Figure 6–2a, then there is enhanced evidence for the presence of an obstacle in the area of intersection. Conversely, in Figure 6–2b, two of the sensors indicate the absence of obstacles, indicating an enhanced belief in the availability of freespace in the intersection area.

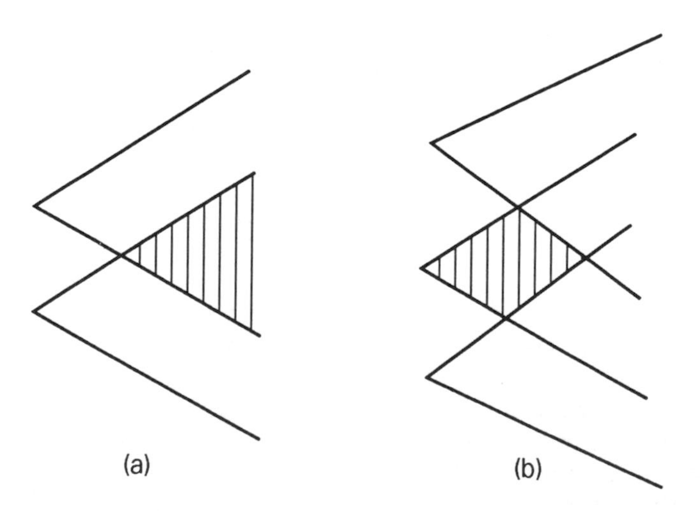


**FIGURE 6–2.** Simple sensor fusion examples: (*a*) enhanced belief; (*b*) restricted belief.

These simple situations are easily handled by the belief mechanisms described in the previous section. Table 6–1 shows the results of strong detection by both detectors, strong detection by one unit and not by the other, and strong detection from neither unit. In the first case, the overlapped region and the total coverage area indicate enhanced detection beliefs of 0.56 and 0.94, respectively, indicating the most likely regions for the obstacle. In the second case, the intersection region has a normalized belief of 0.16, the total coverage area 0.65, and the indeterminacy or ignorance 0.11. In the third case the bulk of the evidence is split between indeterminacy and belief in freespace, with only slight beliefs of 0.01 and 0.14 of obstacles in the regions of overlap or total coverage. Thus the belief function results correspond to our intuitive expectations for these simple cases.

In order to combine the belief functions from multiple sensors, it is necessary to perform a multidimensional Dempster direct product. All intersections must be generated and corresponding cross products computed and accumulated. The normalization factor is computed by summing the cross product components which result in null intersections, and subtracting this sum from unity. The belief function for any region can be computed by summing all contained components and normalizing. Efficient data structures and algorithms for computing the direct product and extracting belief function values are discussed by Graham and Smith (1988).

**TABLE 6–1 Simulation Results for Two Sensors**

| | Case I | Case II | Case III |
|---|---|---|---|
| Sensor 0 | | | |
| Detection | 0.8 | 0.1 | 0.1 |
| Nondetection | 0.0 | 0.6 | 0.6 |
| Uncertainty | 0.2 | 0.3 | 0.3 |
| Sensor 1 | | | |
| Detection | 0.7 | 0.8 | 0.1 |
| Nondetection | 0.0 | 0.0 | 0.0 |
| Uncertainty | 0.3 | 0.2 | 0.9 |
| Normalized weights | | | |
| Region A | 0.56 | 0.16 | 0.01 |
| Region B | 0.24 | 0.04 | 0.10 |
| Region C | 0.14 | 0.48 | 0.03 |
| Belief functions | | | |
| Obstable in intersection | 0.56 | 0.16 | 0.01 |
| Obstacle in coverage area | 0.94 | 0.65 | 0.13 |
| No obstacle | 0.00 | 0.23 | 0.57 |
| Unresolved uncertainty | 0.06 | 0.11 | 0.29 |

Figure 6–3 shows a sensing system with more sensors and more sensing regions. The decision task is to determine if the robot can move within the quadrant. Sensor 0 is a beam-type sensor with two geometrically separated detection regions, such as represented by the "fingers" of an infrared detector. Sensor 1 is a beam-type sensor with a strong detection within a relatively small region of the beam. This is a typical detection pattern for an ultrasound sensor. Sensor 2 in an area-type sensor with two adjacent detection regions. Sensor 3 reports no detection within the region. Table 6–2 indicates the basic belief

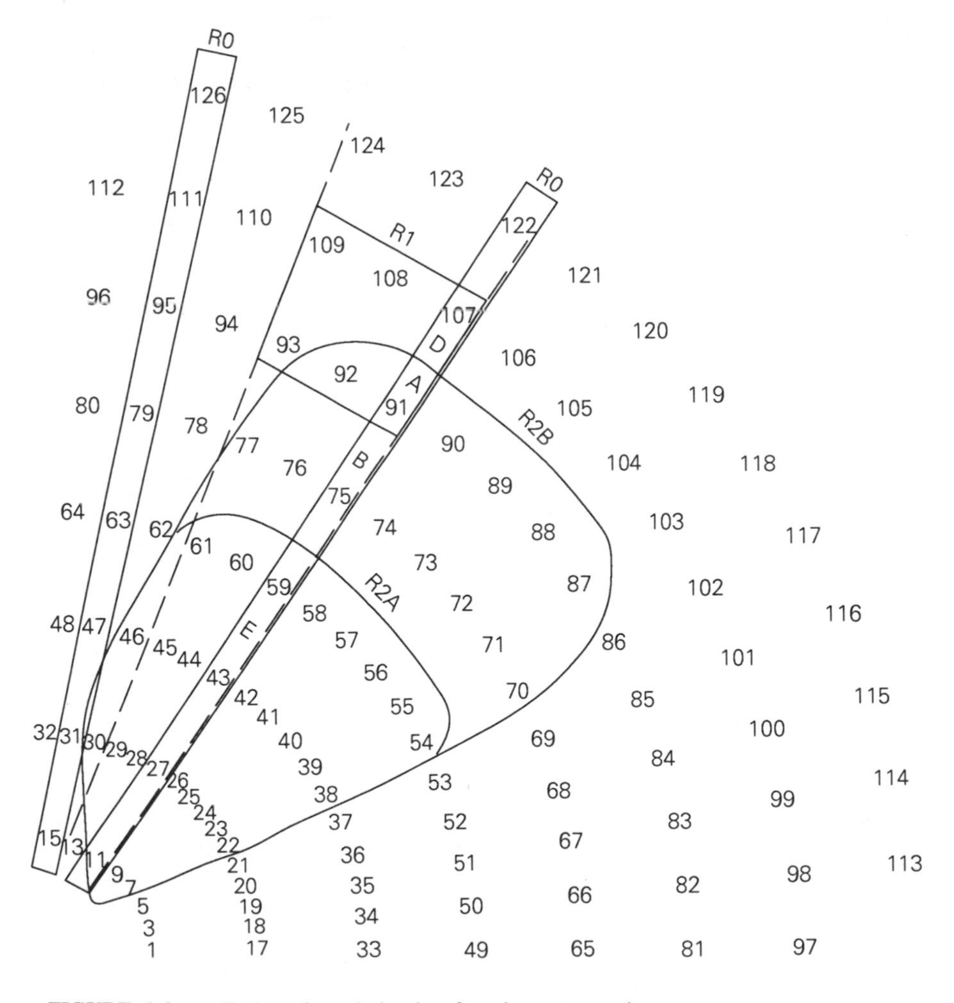


**FIGURE 6–3.** Fusion of proximity data from beam-type and area-type sensors.

**TABLE 6–2 Simulation Results for Multiple Sensors**

| | Probability | Indeterminacy | Belief |
|---|---|---|---|
| Sensor 0 | | | |
| Region 0 | 0.8 | 0.2 | 0.77 |
| Sensor 1 | | | |
| Region 1 | 0.9 | 0.1 | 0.89 |
| Sensor 2 | | | |
| Region 2 | 0.7 | 0.3 | 0.69 |
| Sensor 3 | | | |
| No detection | 0.5 | 0.5 | 0.006 |
| Normalization factor = 0.51 | | | |

assignments associated with each sensor and the belief functions after processing. Point 91 on the grid accumulated a belief of 0.50, which is reasonable since it is the intersection of three regions of detection. Any region containing this point will thus indicate a strong belief in the presence of an obstacle. The total belief of an intruder in the scanned region is 0.94.

A limitation of the system above is the failure to utilize fully information about regions where sensors do not detect obstacles (as opposed to rather detailed information about the presence of obstacles). This situation can be addressed by generating a second belief structure for the absence of obstacles (i.e., the presence of freespace).

## 5. THE DECISION LEVEL

The decision level is the highest level of the control hierarchy and has the responsibility for using the information from the workplace map to determine admissible operating conditions for the robot system. By the hierarchical structuring principle of Saridis (1979), this level should operate with the greatest degree of intelligence but the least amount of precise detail. (This effective reduction in the decision space granularity is referred to by some investigators as multiresolution control.) In many cases there are no clearly defined criteria that delimit the scope of function for the uppermost level of a multiresolution controller. Fortunately in the case of the hierarchical robot safety controller for industrial robots, the set of outputs is a collection of a small, finite number of well-defined alarm states, thus simplifying the problem.

Formally, let $\Omega = \{o_1, o_2, ..., o_n\}$ be the (finite) set of allowable output states for the safety controller. Then the fused sensory data from the previous level map onto these outputs as follows:

$$D \rightarrow \Omega, \tag{6–8}$$

where $D$ is the decision space mapping of the sensory fusion function as discussed in the previous section. Thus, there are two issues that must be resolved to complete the hierarchical design. The first is the selection of a performance index for partitioning the decision space among the outputs, and the second is the question of an efficient implementation of the decision level controller.

Considering the second issue first, this research effort has taken the approach that an object-oriented, knowledge-based representation of the problem will provide a superior long-term solution, although possibly suffering inefficiencies on conventional hardware. The advantage of this approach is that the logical

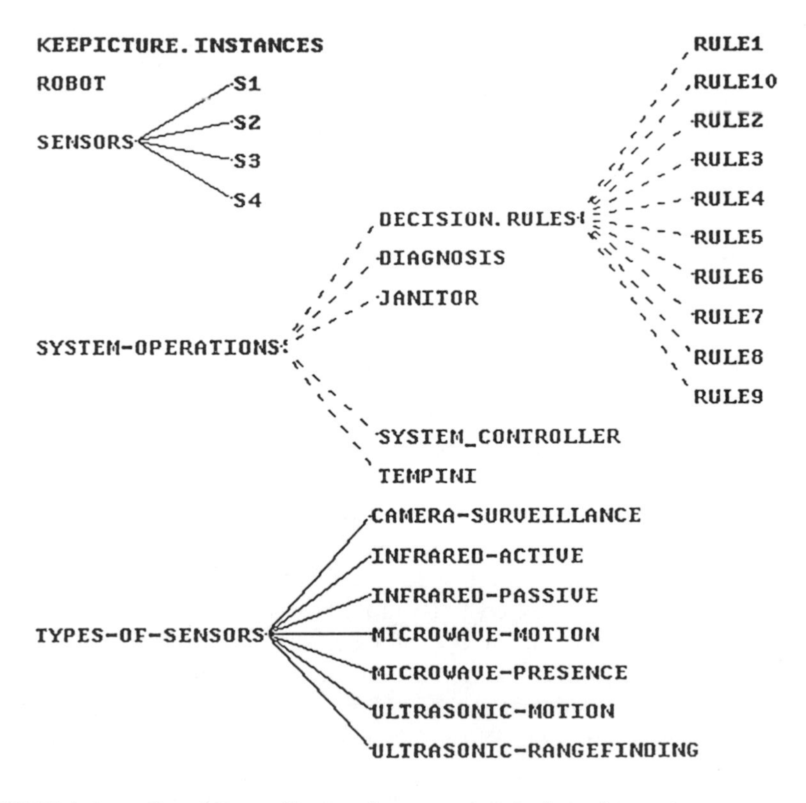


**FIGURE 6–4.** Knowledge architecture for sensory-based robot safety system.

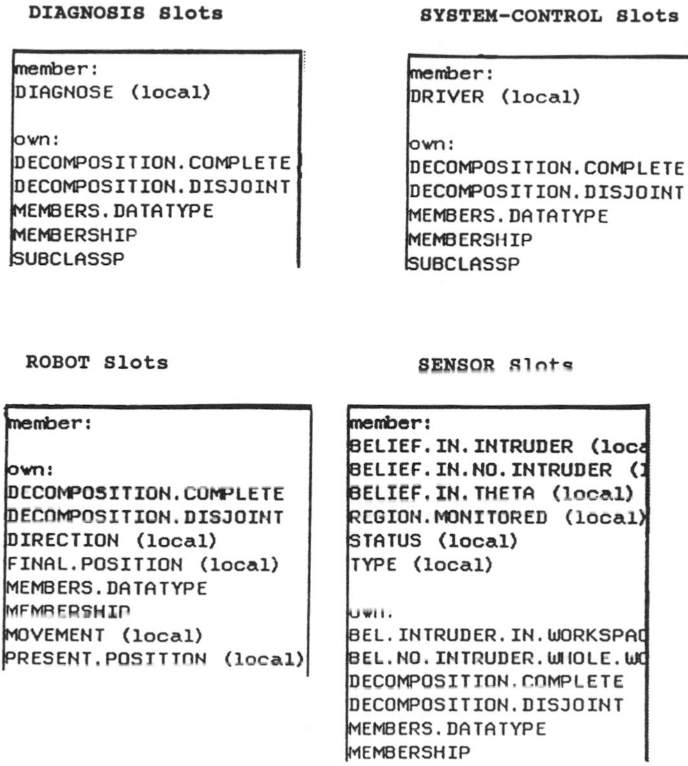

**DIAGNOSIS Slots**

```
member:
DIAGNOSE (local)

own:
DECOMPOSITION.COMPLETE
DECOMPOSITION.DISJOINT
MEMBERS.DATATYPE
MEMBERSHIP
SUBCLASSP
```

**SYSTEM-CONTROL Slots**

```
member:
DRIVER (local)

own:
DECOMPOSITION.COMPLETE
DECOMPOSITION.DISJOINT
MEMBERS.DATATYPE
MEMBERSHIP
SUBCLASSP
```

**ROBOT Slots**

```
member:

own:
DECOMPOSITION.COMPLETE
DECOMPOSITION.DISJOINT
DIRECTION (local)
FINAL.POSITION (local)
MEMBERS.DATATYPE
MEMBERSHIP
MOVEMENT (local)
PRESENT.POSITION (local)
```

**SENSOR Slots**

```
member:
BELIEF.IN.INTRUDER (loca
BELIEF.IN.NO.INTRUDER (l
BELIEF.IN.THETA (local)
REGION.MONITORED (local)
STATUS (local)
TYPE (local)

own:
BEL.INTRUDER.IN.WORKSPAC
BEL.NO.INTRUDER.WHOLE.WO
DECOMPOSITION.COMPLETE
DECOMPOSITION.DISJOINT
MEMBERS.DATATYPE
MEMBERSHIP
```

**FIGURE 6–5.** Unit definitions for safety knowledge base.

relationships among the various system entities can be captured by a well-defined inheritance network, independent of the details of the sensory network and preprocessing system, and then the representation can be customized to the specific configuration of any safety system by adding the correct number of instance modules. Also a set of production rules can be generated to manipulate the slot values, which is again independent of specific implementation details.

Figure 6–4 shows the preliminary architecture of a knowledge base for the decision making function of a sensory-based robot safety system. In its present form the knowledge base consists of seven units representing the basic elements of the safety system, as shown in Figure 6–5, and ten rules that define the appropriate system responses. The structure of the knowledge base has been intentionally kept simple for feasibility evaluation, with the realization that more rules would be used in an actual system. This knowledge base is presently implemented in Intellicorp's KEE Knowledge Engineering Environment. Current response time of the system is on the order of 1–2 sec, including quite a bit of screen output. Although this is two orders of magnitude slower than what will be required for real-time operation, we estimate that a dedicated run-time version of the system with currently available hardware could achieve real-time performance for a system with 10 or fewer sensors.

## 6. CONCLUSIONS

This chapter has presented one approach to the unified design of a sensory-based robotics safety system for the current generation of fixed location industrial robots. The key to a robust and reliable design is to distinguish between the functional requirements and the implementation constraints that limit performance of such a system. This paper presents a three-level hierarchical design that encompasses both sensory processing and fusion of sensory data, and the high-level decision making processes for a graduated alarming scheme.

Clearly there are a number of limitations in the present design in terms of the computational requirements for the fusion and higher-level decision functions. However, recent research on specialized computer architectures, and the continuing evolution of microelectronics make these features seem achievable within the foreseeable future.

## ACKNOWLEDGMENTS

This work was supported in part by grants from the Kentucky EPSCOR program and from the University of Kentucky Center for Robotics and Manufacturing Systems.

### References

Albus, J. 1981. *Brains, Behaviour, and Robotics*. New York: McGraw-Hill.

Bennekers, B. and Ramirez, C. 1986. Robot obstacle avoidance using a camera and a 3-D laser scanner, SME Vision 86, Detroit, MI. pp. 2.57–2.65.

Dempster, A. 1967. Upper and lower probabilities induced by a multivalued mapping. *Annals of Mathematical Statistics* (38):325–339.

Graham, J., et al. 1986. A safety and collision avoidance system for industrial robots. *IEEE Trans. Industrial Applications* IA-22:195–203.

Graham, J., & Smith, P. 1988. Computational considerations for robot sensory integration. Proc. Symposium on Advanced Manufacturing, Lexington, KY, pp. 115–118.

Kilmer, R. 1982. Safety sensor systems for industrial robots. *Proceedings of the SME Robots 6 Conference,* pp. 479–491.

Kilmer, R., et al. 1984. Watchdog safety computer design and implementation. *Proceedings SME Robots* 8.

Meagher, J., et al. 1983. Robot safety and collision avoidance. *Professional Safety* 28:14–18.

Saridis, G. 1979. Toward the realization of intelligent controls. *Proceedings of the IEEE* 67(8):1115–1133.

Sneckenberger, J., Kittiampton, K., and Collins, J. 1987. Interfacing safety sensors to industrial robotic workstations. *Sensors* April:35–37.

Shafer, G. 1976. *A Mathematical Theory of Evidence*. Princeton, NJ: Princeton Univ. Press.

Sugimoto, N. and Kawaguchi, K. 1983. Fault tree analysis of hazards created by robots. 13th Intl. Symp. Industrial Robots. pp. 9.13–9.28.

Zadeh, L. 1978. Fuzzy sets as a basis for a theory of possibility. *Fuzzy Sets and Systems* 1(1):3–28.

# 7

# An Intelligent Safety System for Robots and Automatic Machines

Heikki Koivo, Timo Malm, Jyrki Suominen, and Risto Kuivanen

## 1. INTRODUCTION

The number of robots and automatic machines in industry increases all the time. Traditional safety devices are often inflexible, and the safety provided by them insufficient. Typically, such systems always stop the machine when a person enters the static danger zone. The zone is usually surrounded by a fence, light curtain, or some other sensor.

With intelligent safety systems, the danger zone need not be static. The shape of the danger zone depends on the location and the task of the machine. Both movement of the machine and human being have to be observed separately.

The central issue in safety systems is the observation of human beings (Graham and Meagher, 1985). Touch sensors represent the classical safety technology. The objectives of an intelligent safety system are easier to achieve by using sensors for detecting the presence of human beings (Irwin and Caughman, 1985). Unfortunately, most of the sensors currently available are not designed for safety applications.

The principles of intelligent safety systems are discussed first. Then the basic components of such a system are reviewed. Finally, two prototype systems are described: one for a welding robot and the other for an automatic stretch film wrapping machine.

## 2. PRINCIPLES OF INTELLIGENT SAFETY SYSTEMS

### 2–1. Safety Systems Applied to Production

The basic principle is that a human being must have a safe workplace at the same time that production needs are satisfied. This should be emphasized when planning the production system.

The main characteristics of an intelligent safety system are as follows:

- Worker is protected during automatic, setting, disturbance, and repair modes.
- It is possible to work safely near the machine, disturbing production as little as possible.
- Production can be started quickly and easily after a stop command by the safety system.
- The safety system is reliable and complies with regulations.
- If there are many machines, a decision must be made whether an alarm in one causes actions in others.

## 2–2. Tasks of a Safety System

In order to improve the flexibility of automatic machines, there must be other stopping modes than just emergency stops. The following modes have been suggested in Linger (1987):

**Production Hold:** The machine is programmed so that it completes the duty cycle or the remaining part of it after receiving a production hold command. After this, it is transferred into a pause state. Restarting is done either automatically or manually. The production hold command is given by either the operator or the safety system. The safety hold command and the emergency stop override the production hold.

**Safety Hold:** Stop commands coming from safety devices stop machines and devices immediately. If production can be stopped at any time, production and safety commands can be combined. The safety hold turns the power to a machine off and prevents all dangerous functions with brakes or locking devices, if needed. The emergency stop overrides the safety hold.

**Emergency Stop:** Emergency stops should only be used in real danger situations, for example, when somebody is in danger of being hurt in spite of the safety system or if the control system becomes faulty. The emergency stop usually stops power delivery to all units, except to those units the stopping of which might cause a danger situation.

Restarting after an emergency stop is usually slow. The start must often be made from the beginning of a duty cycle, which is troublesome from the production viewpoint. An emergency stop can be performed by an operator or a self-acting emergency device, like a monitoring circuit in a safety system.

The functioning of a hydraulic robot illustrates the differences between an emergency stop and a safety hold. A safety hold opens the circuits going into hydraulic valves, causing the closing of valves and the stoppage of the robot. In addition to the previous actions, an emergency stop also causes the stoppage of hydraulic machinery, stopping among other things the oil flow under pressure from a broken hydraulic line.

Other modes of operation in a safety system may also be used. One is to slow down the operation speed when a human being is working close to a machine. This increases flexibility in both safety and production.

The safety system should give information about control actions and observations for the worker. This could include light or sound alarms, when a human being comes close or enters the danger zone. The state of different sensors and actuators can be described by LEDs (light emitting diodes) and lamps in a layout panel of the automation system. Alphanumerical displays are easily attached to a programmable logic and natural language statements to a microcomputer system. This, however, is expensive in small systems.

## 2–3. Structure

Although details of a safety system are determined from one application to another, its general structure can be described with a block diagram. A safety system has three main parts: a detection unit, a decision unit, and a control unit (Figure 7–1).

The detection unit consists of sensors, control buttons and switches, and other devices, which control the operation of the decision unit. The information collected from the protected system and environment can be:

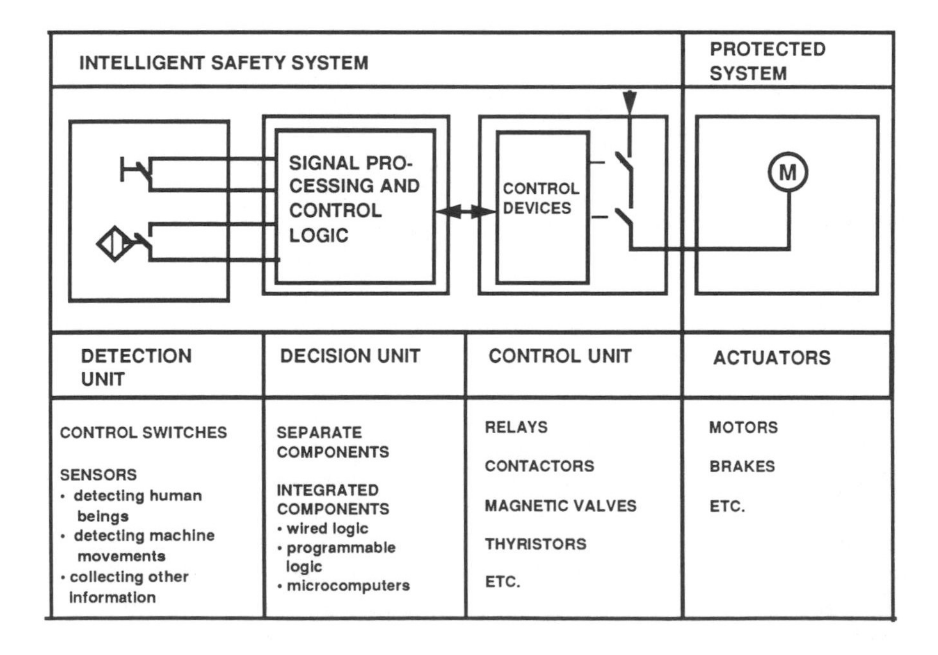


**FIGURE 7–1.** General structure of a safety system.

- Position and direction of movements of human being.
- Followup of movements of automatic machine.
- Duration of actions.
- Logic of sensor information.
- Monitoring position of workpiece.
- Use of frequency counters, which are able to detect rotation or changes in rotation speed.
- Force measurements, which detect if workpiece is stuck or if robot has hit something.

A fault in the detection unit must not lead to a hazardous situation. This can be ensured either by using failsafe components or by redundant components.

Signals from the detection unit are processed in the decision unit. Signal processing must be ensured, that is, the protective effect must not be lost because of a fault. Traditional rely logic can be added to a microcomputer system. The primary target is to make the microcomputer control failsafe. Another possibility is to use parallel microcomputer systems and then compare their outputs. If they are not equal, the protected system is stopped.

The control unit can consist of relays, contractors, magnetic valves, etc. Standard BS 6491 (1984) determines that the device must have at least two switching devices, which are monitored and which affect machine movement directly. According to the same standard, if a fault is detected in a part that is monitored, the safety device must move into a safe state.

## 3. STRUCTURAL COMPONENTS OF AN INTELLIGENT SAFETY SYSTEM

### 3–1. Sensors Detecting Human Beings

Human beings can be detected by positioning sensors on a robot arm or its surroundings. If a sensor is put into the robot arm, a safe three-dimensional zone can be generated around the robot. If the sensor is placed into the surrounding area, the danger zone is often larger. This also holds true for automatic machines. An example of this situation is a portal robot. The overall issue is always reliability. Detectors can be divided into the following classes:

- Touch detectors
- Optical sensors
- Passive infrared detectors
- Ultrasound sensors
- Microwave sensors
- Capacitive sensors
- Vision systems

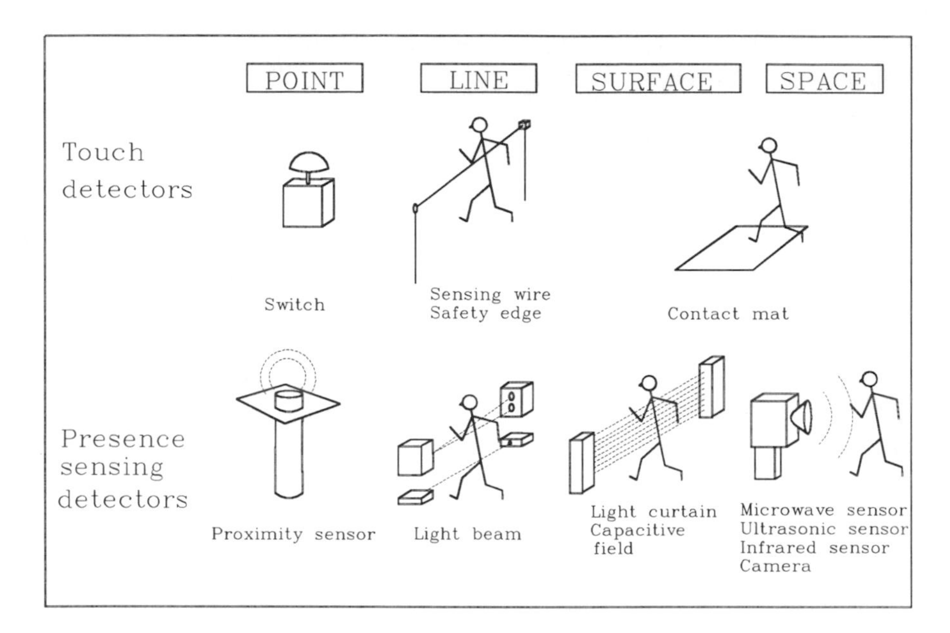


**FIGURE 7–2.** Principles of touch and presence sensing detectors.

Touch sensors are suitable for monitoring pointlike and linelike objects and surfaces. Contact mats are reliable and suitable for a robot, but would demand an extensive area to be covered for an automatic machine. Contact mats are usually based either on pneumatics or electrohydraulics.

A pneumatic contact mat has a pressure sensor and a receiver. A pipeline between the two has been positioned so that no message is received when somebody steps on the mat. Pneumatic mats have a fairly slow response time and are quite expensive.

Electrohydraulic contact mats have two conducting metal strips and between these flexible material. In Sweden, the space between the strips must be so small that a cylinder of 10 kg with a 25 cm diameter can be detected (AFS, 1980). Figure 7–2 illustrates touch- and presence-sensing detectors.

## 4. INTELLIGENT SENSORS FOR ROBOTS

The sensor actions must be such that the robot or its peripheral devices cannot harm a human being. On the other hand, the sensor should not slow down work unnecessarily.

An intelligent sensor for robots consists of a decision unit, sensors for following the robot, and sensors for following human beings. The decision unit receives information from different sensors. If a human being approaches the robot, the decision unit gives the robot a safety speed, or stop command. The command depends on the distance of the human being, the status of the push button, and the state of the robot. If a dangerous fault is detected, the decision unit gives an emergency stop command.

Next a prototype safety system is described. It is designed for a robot cell, which consists of a Motoman L10W welding robot, two welding benches, and a jointed jig (Figure 7–3). It can be used with other robots as well.

### 4–1. Decision unit

The main parts in the decision unit are a programmable logic (or a microcomputer) and relay switches. Either of these can give, independent of each other, a stop or safety speed command. (In special cases the programmable logic may also give an emergency stop command.) Programmable logic and relay logic operate simultaneously, because then a delay in detecting fault in one logic does not create dangerous situations.

In principle, a fault in either programmable logic or relay logic will be revealed. This will take some time, e.g., in the programmable logic about 0.15

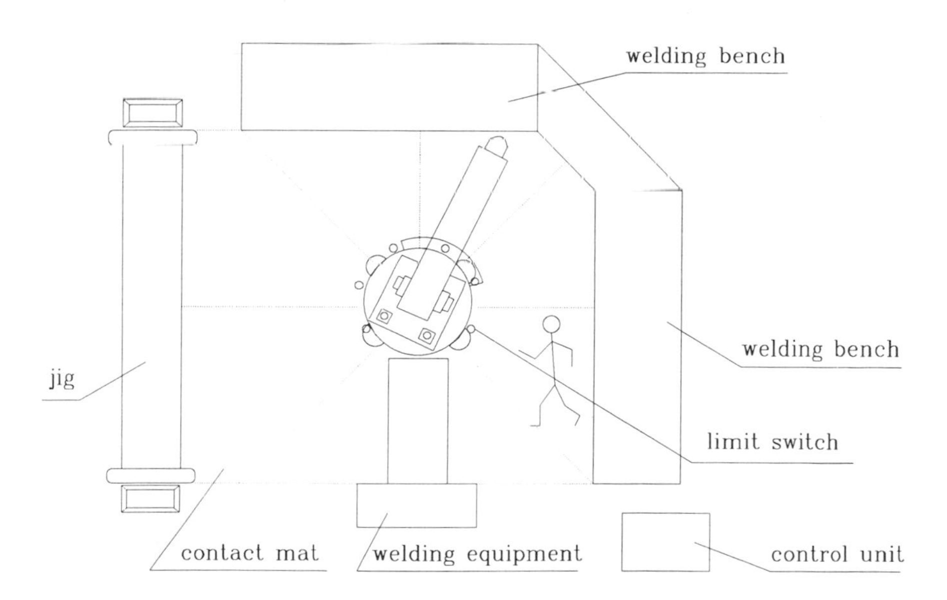


**FIGURE 7–3.** Schema of the designed intelligent safety system for industrial robots.

sec. In the relay logic, if sufficiently many relays are out of order, only a human being can detect the fault by "observation from the danger zone" lamp.

The relay in stop and safety speed loops controlled by the programmable logic is connected in series with the relay switches of relay logic. Programmable logic alone takes care of the emergency stop and its watchdog circuit.

A practical aspect that must be kept in mind is a steady supply of electricity. Programmable logic requires 24 V dc. If any known voltage disturbances exist, separate disturbance protection circuits might be required, although they are relatively expensive. Voltage peaks must also be protected against.

### *Relay Logic*

Relay logic is usually very reliable. Faults start occurring more frequently in older units. Their reliability function follows a bathtub curve. Although rare, it is worthwhile to take relay faults into consideration. These are detected by a watchdog circuit.

The operation of relay logic is fairly simple (Figure 7–4). A **safety speed** command is given when:

- Human being is on the contact mat
- Push button is in safety speed position

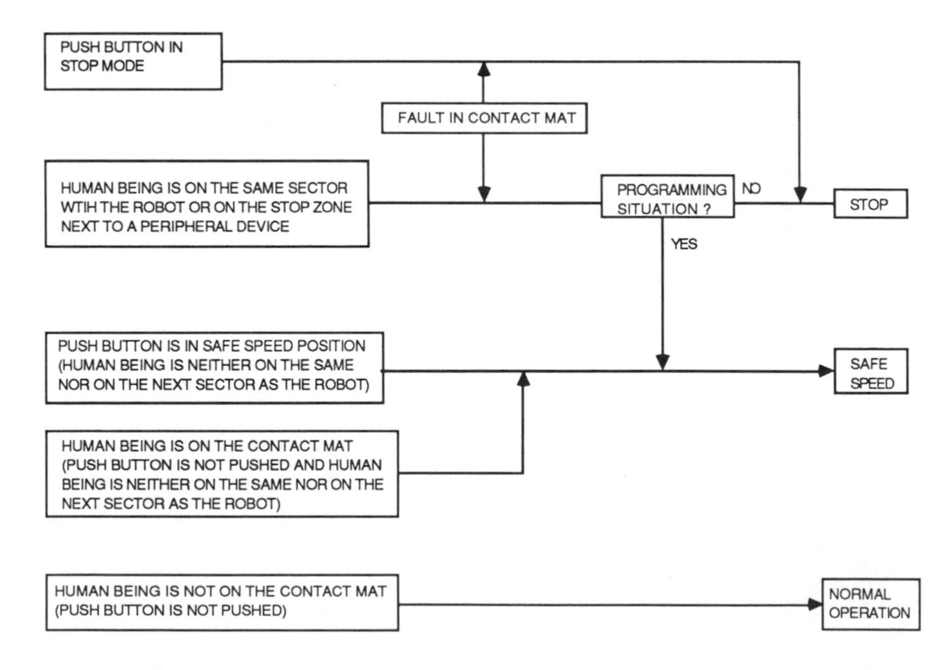


**FIGURE 7–4.** Operation of relay logic.

Relay logic gives a **stop command** when:

- Human being and robot are in the same sector, not during programming stage
- Human being is on the contact mat causing the stop
- Push button is in stop position

***Programmable Logic***

Programmable logic receives sensor information from contact mats, robot following, push button position, and robot status information. It can give safe speed, stop, or emergency stop commands, and direct lamps. The lamps give information about two things: human being on contact mat or sensor fault.

Based on the status information and the position of the push button, the programmable logic decides how near the human being can get without causing the robot to stop. It also monitors the operation of the stop and safe speed commands based on status information from the robot. If the robot did not obey the safety speed command, the logic would give a stop command when the human being goes near the robot. If the stop command did not function, logic would give an emergency stop command. In this way also the robot functions can be checked.

The following tasks are taken care of by the programmable logic and not by the relay logic:

- Programmable logic gives a stop command, if the human being is in the next sector to the robot and push button is not pushed. This becomes active only after 0.5 sec has elapsed since moving into safe speed. Such an interval is very short for human actions, but the decision unit is able to check the functioning of the robot and the robot speed has time to settle to a safe level.
- Programmable logic keeps information "human being on contact mat" for 2 sec after the person has left. This ensures that the robot knows the position of the human being, although there is no sensor information, in case that the person jumps or runs.
- Programmable logic holds safety speed and stoppage for about 3 sec after the need for stoppage or safe speed. This prevents a sudden start-up of a robot before the human being has left the danger zone.
- Programmable logic checks if the robot is in stop mode 3–4 sec after the command has been given. If the robot is not in the stop mode, it receives an emergency stop command.
- Programmable logic does not check the execution of the safety speed command based on time, because it varies a lot. The assurance of the safety speed is realized so that the human being is allowed into the sector next to the robot arm only when the push button is pushed and the safety speed is in operation.

In case the robot does not constantly inform the logic of being in the safety speed mode, the speed is checked by monitoring the changes in the robot state.

- Programmable logic checks if the position information provided by the limiting switches is logical. This function monitors the correct operation sequence of relays and observes the nonoperable limiting switches.
- Programmable logic keeps the observed fault situation in memory and stops the robot when a dangerous situation develops. If a fault is detected in the limiting switches, the robot position information is not certain anymore and a stop command is given in case a human being enters the danger zone. A fault in the contact mat corresponds to the situation: human being on contact mat. Dangerous relay faults hold the robot in stop mode.

## 4–2. Sensors Used

Mechanical limit switches are used to determine the position of the robot. These can be positioned at the intervals of 30° on the rack positioned in the robot base. A metal cam is attached to the robot. The cam is on top of the rack containing the limiting switches. It is in contact with at least the limiting switch, corresponding to the sector, where the robot arm is located.

Contact mats are used to detect human beings. A push button is useful to ensure further safety. The use of a push button has the following advantages:

- When pushing a push button, a human being acknowledges that he or she is going into the danger zone.
- A human being has a chance to stop the robot by pushing the button. When the push button is in the safety speed position, the human being can stop the robot by pushing the button all the way to the bottom.
- A human being can give the safety speed command to the robot in time. When a robot is moving rapidly, its transfer into the safety speed mode is relatively slow.

A human being is quite slow in reacting to danger. Therefore, a push button alone is not enough. Indicator lamps are used to show the status of the decision unit. If an indication lamp is on, the decision unit has detected a fault in itself and gives a stop command in case of danger.

The overall operation of the intelligent sensor is displayed in Figure 7–5. The diagram of the intelligent sensor implemented is displayed in Figure 7–3. The system was installed in an industrial site in 1987 and has been operating there ever since. Figure 7–6 shows the overall industrial installation and Figure 7–7 shows the contact mat around the robot. There has been no major problems with the safety system. In practice the use of the push button has been tedious. This is

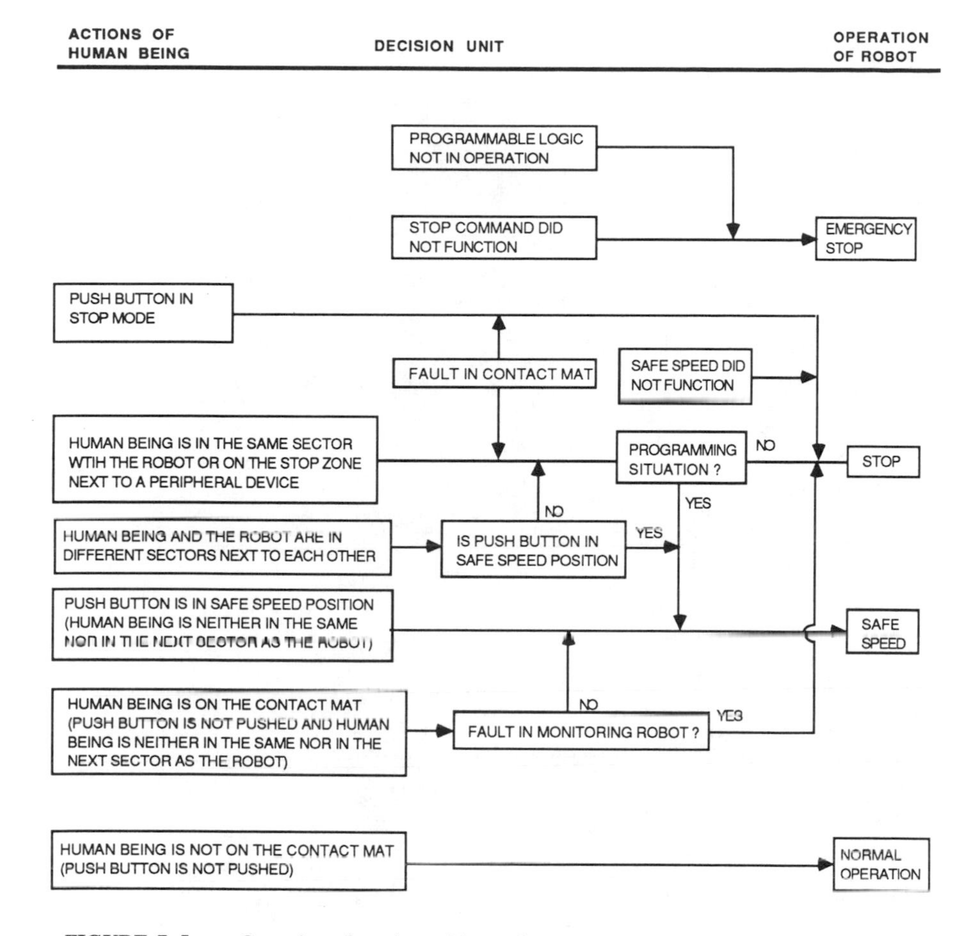


**FIGURE 7–5.** Operation of a robot with a safety system.

perhaps the reason that there has been a tendency to turn the intelligent sensor off.

## 5. INTELLIGENT SENSOR FOR AN AUTOMATIC MACHINE

A safety system for an automatic wrapping machine was considered. The material to be wrapped is piled on a packing platform and brought into the machine with a conveyor belt. The entrance of the platform is observed by a photosensor. If the machine is empty, the platform is allowed into the wrapping machine and is centered there with the aid of a photosensor beam. After this the

**FIGURE 7–6.** The overall view of the installation.

**FIGURE 7–7.** Contact mat around the robot.

wrapping frame is lowered around the platform. The foil roll rotates together with the frame around the platform and moves upward at the same time. When the wrapping has been finished, the foil is cut with hot wire and the end of the foil is heat-sealed. The package is now ready and the conveyor belt takes it out. The procedure is then repeated.

The control of the machine is realized with programmable logic. The machine can be stopped by stop and emergency stop push buttons. There is no need for a separate production stop, since the machine can be stopped and restarted at any phase of wrapping.

The wrapping machine can be protected by putting it into a cage. The cage can, however, be replaced by an intelligent safety system, which also maximizes safety by protecting the input and output entrances of the platform.

The principal danger element in the machine is the rotating wrapping frame and the attached foil roll rack. The heat sealer can cause a burn, but since the time interval of operation is only a few seconds at the end of the wrapping, this situation is not very likely. The danger zone is thus a connected space, limited by the frame of the wrapping machine.

## 5–1. Sensoring Solutions

Since sensors are not able to limit the movements of human beings into the danger zone, a zone outside the wrapping machine must be monitored. This zone is such that during the time it takes to cross over it, the movement of the dangerous machine can be stopped. Sensoring must also be able to differentiate the platform entering or leaving the machine from a human being.

The outside area of the automatic machine is covered by infrared detectors and photosensors. Infrared sensors do not yet meet all the requirements of safety devices as far as reliability is concerned. For the time being the detectors are duplicated. One detector can only cover one side. Because of the duplication, altogether eight detectors are required. They are mounted in the middle of each side. Since human beings cannot see the detection field of the infrared sensor, the monitoring zone should be clearly marked on the floor.

Infrared sensors are sensitive only to human beings—platforms on the conveyor do not cause alarms. Based on the sensor information, the speed of the wrapping frame is lowered.

Photosensors are mirror-deflecting with a maximum operating distance of 8 m. The sensor has a monitoring circuit, which checks the dirt on the lenses. Since a photosensor itself does not have monitoring required in safety devices, it must be duplicated. Only crossing the photosensor beam results in machine stoppage. On the conveyor side this information does not reveal whether it is a human being or a platform that crosses the beam. For this purpose an additional photosensor is needed.

At the exit the decision procedure is however easy, since the platform and the human being travel in opposite directions. The extra photosensor is installed closer to the frame. The outcoming platform first crosses the inner and then the outer light beam. The human being, on the other hand, first crosses the outer beam. In such a case, the machine is stopped.

At the entrance, the situation is somewhat more delicate, since both the platform and the human being travel to the same direction. Therefore, an extra photosensor is installed outside the actual stopping line. The entering object is identified as a platform when:

- The conveyor belt is operating.
- The outer photosensor gives the signal first, and the signal is continuous.
- The time difference between the inner and outer photosensors is correct. Since the conveyor moves at a constant speed, the time difference is known.

If any of these conditions is not fulfilled, the object is a human being.

The inside of the wrapping machine is not fully monitored. This would be a rather difficult task. Contact mats cannot be used efficiently, since the wrapping machine covers a lot of the floor space. Other sensor solutions are also of limited use because of the platform and moving machine parts. The only sensor installed inside the wrapping machine is a mechanical safety edge in front of the foil roll rack. This is to certify the unlikely cases that:

- Human being has somehow been able to enter the danger zone and is in front of the foil roll rack movement.
- Machine is restarted when human being is still inside the danger zone. Since the inside of the machine can be seen by the machine operator and the stoppage must be acknowledged, this is rather unlikely.
- The platform is on the route of the foil roll rack. This is possible if, e.g., the centering has failed due to a sensor failure.

## 5–2. Decision Unit

Here again a programmable logic is used as the decision unit. For economic reasons it does not make sense to duplicate programmable logic, relay logic is used instead to certify safe operation. Their operation can be independent. Their operation must be simultaneous so that a delay in observing a fault does not create a dangerous situation.

### *Relay Logic*

Relay logic receives the following data: information from all sensors, acknowledgment information about slower speed, and information from the wrapping machine–conveyor in operation and wrapping frame in full speed.

Relay logic allows **normal operation** if:

- Human being is not in the observation zone of sensors, and the sensors, decision unit, and the relays in the current circuit of the wrapping machine are operating properly.

Relay logic gives command for **slow speed** if:

- Wrapping frame is in full speed and a human being enters the monitoring zone of infrared sensors.

Relay logic gives **stop command** if:

- Human being crosses the observation beam of the photosensors.
- The limit switch inside the wrapping machine on the wrapping frame gives an observation.
- A fault in relays is detected.
- A fault in programmable logic is detected.

***Programmable Logic***
Programmable logic is responsible for the same functions as the relay logic, and in addition for the following tasks:

- Checks that the pulse count difference of the photosensors at the platform entrance is correct.
- Gives a stop command, if the photosensors, monitoring either the entrance or exit of the platform, and the infrared sensor give a signal. This implies that a human being and a platform are simultaneously entering the machine.
- Checks that, if sensors are duplicated, signals are received from both.
- Operation of photosensors and infrared sensors are checked after a stop command.

## 5–3. Operation of Safety System

Normal operation occurs when no human beings are observed in the sensor observation zone and the self-diagnosis system has not detected faults in sensors, decision unit, or relays in the current circuits of the wrapping machine. Operation of the safety system is summarized in the block diagram of Figure 7–8.

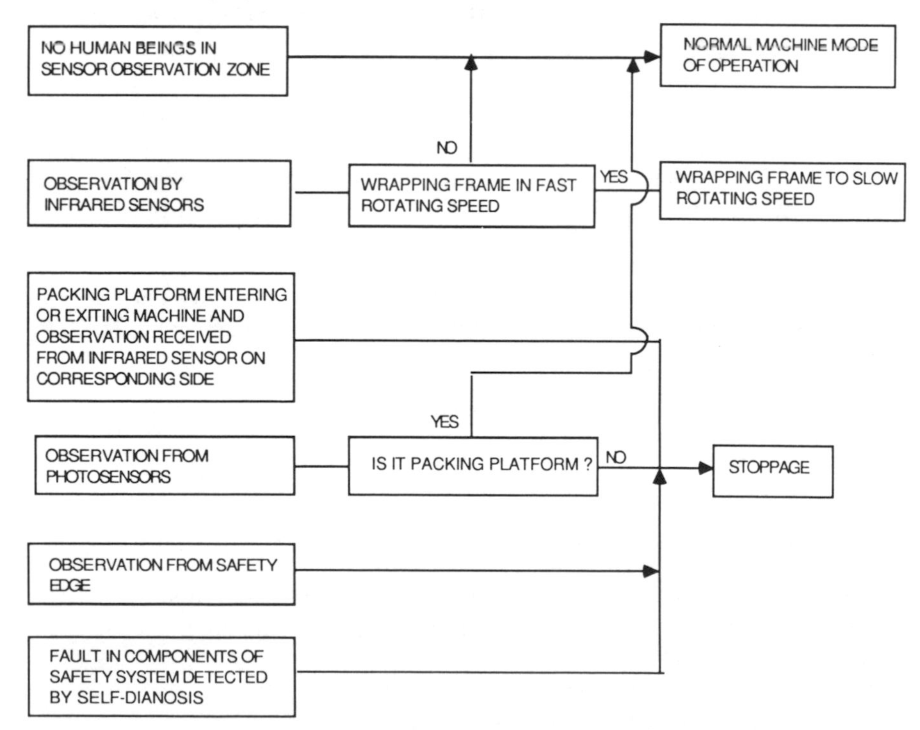


**FIGURE 7–8.** Operation of the safety system of the wrapping machine.

In this case the system has not been installed. The reason was not technical but rather economic. The development funds for the project ran out before its installation.

## 6. CONCLUSIONS

The object of this research has been to demonstrate by two example systems that reliable and economically feasible intelligent safety systems for automatic machines including robot manipulators can be designed and implemented.

The greatest problem in the case of the wrapping machine was to find a solution to the sensor problem. Although this seems easy, it is the large area to be covered that makes the problem difficult. In an economic solution only a limited use of touch sensors is available. Since there are no such sensors that become safely defective at the present, a straightforward solution is to duplicate both photosensors and passive infrared sensors. This increases the cost of the system.

The suggested safety system effectively protects human beings. It can differentiate a platform from a human being and protect areas that have previously been left unprotected.

**References**

AFS 1980:9. *A directive of worker protection agency about electric contact mats with control devices* (in Swedish). Swedish Statute since Jan. 1. 1982. Stockholm.

BS 6491: Part 1: 1984. *Electro sensitive-safety systems for industrial machines.* British Standards Institute.

Derby, S., Graham, J., and Meagher, J. 1985. A robot safety and collision avoidance controller. In *International Trends in Manufacturing Technology. Robot Safety.* eds. M. Bonney and Y. F. Young. pp. 237–246. Berlin: Springer-Verlag.

Graham, J. H., and Meagher, J. H. 1985. A sensory-based robotic safety system. *Proceedings of IEE* 132(4):183–189.

Irwin, C. T., and Caughman, D. O. 1985. Intelligent robotic integrated ultrasonic system. *SME Technical Paper* MS 85-620, pp. 19–39, Dearborn, MI: SME.

Linger, M. 1987. Are robots safe? *Robot Safety.* Gothenborg Sweden: Institutet för Verkstadsteknisk Forskning.

# 8

# A Study of Worker Intrusion into Robot Work Envelope

Waldemar Karwowski, Hamid R. Parsaei, Bangalore Amarnath, and Mansour Rahimi

## 1. INTRODUCTION

### 1–1. The Effects of Industrial Robotization

The American industry is changing from using traditional machines to automated systems with strong emphasis on robotic workstations. Safety and health professionals are becoming concerned with hazards associated with these nontraditional and complex systems. Providing safe robotic workstations is clearly one of the requirements for effective utilization of this technology to improve industrial productivity. Although in the past, the subject of robot design has been studied from the standpoints of mechanics and kinematics, it must be considered from the human perspective today (Parsons, 1985; Karwowski et al., 1987; Yokomizo et al., 1987).

A robot system consists of three elements: (1) human operator, (2) the industrial robot, and (3) the human–robot interface. Hazards, which can lead to an accident, may occur in any of the three modes of operation, i.e., (1) normal operation, (2) maintenance, and (3) teaching (programming a robot). Studies conducted in Sweden and Japan (Backstrom and Harms-Ringdahl, 1983; Sugimoto and Kawaguchi, 1983) have indicated that most of the robot-related accidents occurred during maintenance, testing, and teaching of the robot. With respect to the part of the body injured, about 30% of the accidents involved the operator's finger, head, or arm, indicating the proximity of operators to the point of operation (Linger, 1988). Another high-frequency category of injury was related to head and neck.

The initial studies on robot-related accidents identified several accident causes and classified those into two categories: (1) engineering (or system design) factors and (2) behavioral and administrative (or organizational system) factors

(Backstrom and Harms-Ringdahl, 1983). From the human factors point of view, the factors of interest here are:

1. Inadvertent robot movement.
2. Failure to stop.
3. Expected and unexpected halts and starts of robots.
4. Inadvertent contact with the start button and other switches.
5. High-speed motion of the robot arm.
6. An individual entering the danger zone of a halted robot for troubleshooting, repair, testing, maintenance, or teaching.
7. Unauthorized entry into the general robot work area.
8. An operator performing workpiece adjustments and positioning.
9. An operator performing a tool change.
10. An operator unfamiliar with a robot's programmed movements.
11. Intentional disabling of safety devices.
12. A robot's work function unknown to the operator.
13. Improper motion path for operator manual tasks.

To date, Japan has had the highest number of fatal accidents due to industrial robots or manipulators. As reported by the U.S. Department of Commerce, Japan has also 62.3% of the robotic world market. According to Nagamachi (1988), the first robot-induced fatal accident occurred in 1978, when the auto loaders crushed the human operator working with a variable sequence robot. Between 1978 and 1987, Japan has reported a total of 10 fatal accidents (Nagamachi, 1986, 1988). The United States reported that the first fatal accident occurred in 1984 (Sanderson et al., 1986) when an experienced worker entered the robot's working area. The worker was pinned between a limiting safety pole and the moving robot's arm.

Jiang and Gainer (1987) conducted a cause-effect analysis on 32 robot-related accidents reported in the United States, West Germany, Sweden, and Japan. Grouping the accidents according to personnel injured, type of injury, and cause of injury, they found that line workers were at the greatest risk of injury followed by maintenance personnel. The pinch-point accidents, caused by a part of the individual's body being pinched between moving parts of the robot, or between a moving part of the robot and nonmoving external fixture, accounted for 56% of all accidents. An impact-related injury with a robot projecting a tool or workpiece in its end effector and striking the individual, accounted for 44% of the accidents. The main cause of injury turned out to be poor workplace design (62%), followed by human error (41%).

The studies reported above indicate that robot-related accidents can happen when workers are present inside, or in close proximity to, the robot's work envelope, and indicate that a common cause of accident can be ascribed to human perceptual and judgmental processes (Carlsson, 1980). In many production environments, the close physical interactions between robots and human operators, which are essential for task completions, may result in high degrees of risk of accidents. For example, according to Sugimoto (1985) the "unnatural" and "unexpected" movements of robots were responsible for about 73% of the robot-related injuries and fatalities. In the reported cases, the operators incorrectly assumed that the robot stopped due to the system malfunction. These accidents involved a person entering the work area when the robot stopped (or was working very slowly) and then unexpectedly started to move again.

## 1–2. Human Factors Engineering of Robotic Workstations

At present, very little is known about how humans react in a high-technology environment and in computerized workplaces surrounded by robots, in which the mobility of machines with risks for accidents is a new characteristic feature (Ayres and Miller, 1983; Mastersson, 1987). Sheridan (1984) suggested that it is necessary to set up a human–robot system to determine experimentally the effective distribution of roles between a robot and a human. Noro and Okada (1983) stressed that human factors can play an important role in optimizing allocation of functions between humans and robots, and in relaxation of physical and mental loads created when humans work alongside the robots.

Human factors engineering in robotics considers how robots and those who work with them interact in the workplace, with emphasis on the design of robot systems, procedures for using them, protection of their users, and the division of labor between robots and human workers (Parsons and Kearsley, 1982). Unfortunately, the human factors research on robot systems, and their effects on safety and well-being of industrial workers, is lagging considerably behind robotics hardware developments. To date, only limited research has been conducted on how workers themselves perceive robots with respect to potential hazards and the job stress they create. Argote et al. (1982) performed a prototype study aimed at investigating how employees, as individuals, perceive and accept the introduction of robotic technology. The results indicate that with experience, workers increase their understanding about what a robot really is. However, with time, workers' beliefs about the potential hazards associated with robots become more complex and pessimistic. Also, there is a significant increase in psychological stress among the workers who interact directly with the robots.

## 1–3. Design and Use of Robotic Systems from Human Perspective

The consideration of human factors in robotics safety should concentrate on the intrinsic limitations of the human perceptual and information-processing systems. The inability of human operators to perceive the hazard associated with periodic robot arm movements is one of the major contributory components to robot-related accidents (Rahimi and Hancock, 1986). As reported by Carlsson (1984), in many robot-related accidents, the operators were quite familiar with their assigned tasks, and had been working with the robots for a period of time before the accidents occurred. The limitations in human perceptual, decision, and action processes suggest the need for design restrictions on robot operational speed and stop–start sequences (Karwowski and Rahimi, 1991).

The above conclusions can be further supported by a new set of rules for design and use of robotic systems proposed by Yokomizo et al. (1987), which point out the importance of human factors considerations. These new rules are stated as follows:

**Rule 1:** Robots must be made and used for the well-being and development of people.
**Rule 2:** Robots must not replace people on jobs people want to do themselves, but must replace people on jobs that they do not wish to do or believe to be hazardous.
**Rule 3:** Robots must be built to such specifications that they do not psychologically or physically oppress people.
**Rule 4:** Robots must follow the command of people so that they do not harm other people, only damaging themselves.
**Rule 5:** If robots replace people on certain jobs, the prior approval of the people affected must be obtained.
**Rule 6:** Robots must be made so that they can be easily operated by people and they can readily perform the role of assistants to people.
**Rule 7:** As soon as robots finish their assigned tasks, they must depart from the place so that they do not interfere with people and other robots.

Rules 3, 4, and 6, which refer to the design requirements for robot systems and safety and well-being of the workers operating robotic workstations, are of particular importance to this research project.

## 2. OBJECTIVES

Robot operators perform an array of important interactive tasks such as programming, trouble-shooting, teaching, maintenance, job set-up, and part feeding within the robot's working envelope. Although many of these tasks are performed outside the robot's work envelope, the operator's attention and safety behavior are still critical to task performance.

Based on the premise that industrial workers may be required to maintain a close proximity to robotic workstations in tasks such as system monitoring, maintenance, programming, trouble-shooting, and teaching (Linger, 1988; Noro and Okada, 1987; Sugimoto, 1985), one of the important concerns with respect to safety is human behavior around the robot systems. In particular, experimental knowledge is needed to describe how workers perceive robot's operational characteristics. This knowledge is one of the main prerequisites for development of effective robot safety standards and prevention of accidents associated with the operation of robotic workstations.

The aim of this study was to investigate the number of unintentional worker intrusions into a robot's working envelope under simulated task operations for two industrial robots. To achieve this goal, one group of subjects (12 industrial workers) was instructed to determine the maximum reach of the robot's arm *(reach zone),* while the other group (12 workers) was asked to determine the closest distance from a robot that they felt safe *(safe zone)*. The assignment of workers into these two groups was done randomly.

# 3. METHODS AND PROCEDURES

## 3–1. Experimental Design

Mixed, three-factorial design (2 × 2 × 2) with the instruction type *(reach zone, safe zone)*, robot size *(small, large)*, and accident exposure *(yes, no)* as main effects, and two nested factors (robot speed and approach angle), were used (see Table 8–1).

**TABLE 8–1 Experimental Design Table; Three-Factorial (2 × 2 × 2) Design with Two Nested Factors (2 × 6: Robot Speed and Approach Angle)**

| Instruction type (IT) | Reach zone | | | | Safe zone | | | |
|---|---|---|---|---|---|---|---|---|
| Robot size (RS) | P50 (small) | | MH33 (large) | | P50 (small) | | MH33 (large) | |
| Robot speed (cm/sec): nested factor | 25 | 90 | 25 | 90 | 25 | 90 | 25 | 90 |
| Approach angle: nested factor | 1 | 1 | 1 | 1 | 1 | 1 | 1 | 1 |
| | 2 | 2 | 2 | 2 | 2 | 2 | 2 | 2 |
| | 3 | 3 | 3 | 3 | 3 | 3 | 3 | 3 |
| | 4 | 4 | 4 | 4 | 4 | 4 | 4 | 4 |
| | 5 | 5 | 5 | 5 | 5 | 5 | 5 | 5 |
| | 6 | 6 | 6 | 6 | 6 | 6 | 6 | 6 |
| Accident exposure (AE)[a] | | | | | | | | |
| Accident group | $s = 3$ | | $s = 3$ | | $s = 3$ | | $s = 3$ | |
| No-accident group | $s = 3$ | | $s = 3$ | | $s = 3$ | | $s = 3$ | |

[a] $s$ = number of subjects randomly assigned to each of the eight (2 × 2 × 2) experimental conditions.

## 3–2. Experimental Layout

The robot was placed in an open area within the laboratory. Lines of approach toward the robot were marked on the floor with a low-contrasting colored tape. A semicircle in front of the robot was equally divided into segments of 25.7° giving six angles of approach. (See Figure 8–1.) The lines were drawn and taped on the floor from the base of the robot to a distance of 335.3 cm beyond its true work envelope. Dummy lines were marked around the robot for uniformity. Safety barriers were placed in an arc at a distance of 365.7 cm. The robot controller was placed outside the barrier at the side of the robot.

The robot's arm was preprogrammed to perform a simulated palletizing task with a speed of 25 cm/sec. The task was done in such a manner that the robot would traverse its entire work envelope encompassed by the six lines. In order for the subject to get a fair idea of the maximum reach in various sectors, the robot was programmed to spend an equal amount of time in each sector, with linear motions covering the minimum and the maximum reach of the robot's arm.

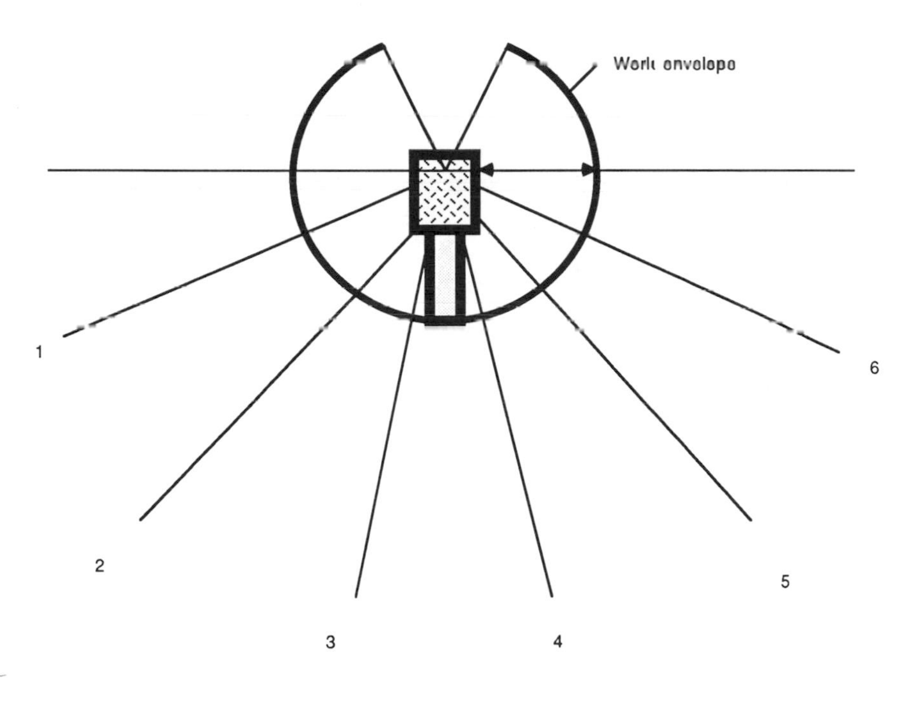


**FIGURE 8–1.** Experimental layout.

## 3–3. Equipment

Two industrial-grade robots were used in the laboratory experiments conducted. One was a "material-handling"-type robot of model MH33 built by Volkswagon Corporation, Germany. The other was a "processing"-type robot of model P50 made by Hitachi Corporation, Japan. Both of the above mentioned robots were built for General Electric Corporation.

The MH33 is a spherical coordinate robot with three degrees of freedom and a servo-controlled drive system. The servo grips have a drift of a 0.25°/min with a repeatability of 0.1 cm. The payload capacity of the robot is 15 kg (33 lb). The arm of the robot is telescopic with a rotation of 320° about the base. The speed of rotation is 59°/sec. The possible tilt of the arm is 30°. This tilt can be achieved at a speed of 50°/sec, whereas the linear motion of the arm is 150 cm. This movement can be covered at a speed of 100 cm/sec. A powered leadthrough programming technique is adopted by the MH33 using a point-to-point playback control system.

The other robot of model P50 is of a jointed-arm configuration with a DC servo-controlled drive system. The repeatability achievable is 1.0204 cm. The range of the five axes being: wrist twist of 370°, wrist bend of 180°, upper arm of 95°, and forearm rotation of 70°. The base rotation possible is 300°. The movement about these axes is with a speed of 100 cm/sec. The maximum reach of this robot is 125.7 cm measured from the tip of the wrist to the axis perpendicular to the base of the robot. The P50 robot is programmed using a powered leadthrough technique utilizing a point-to-point playback control system.

## 3–4. Procedures

Twenty-four industrial workers participated in the experiment. The subjects were randomly divided into two groups of equal size. One group was asked to perceive the maximum reach of the robot's arm *(reach zone),* while the other group was asked to perceive the limits of a *safe zone* around the working robot. During the preview session, half of the subjects in each group were exposed to the simulated industrial accident with the robot speed of 90 cm/sec. During the accident exposure session, the robot arm hit the mannequin placed inside the robot's working space.

In the preview session, the robot ran continuously, randomly touching all points in the sector for a 15 min period. For about 15 min, the subjects were asked to observe the robot on a simulated palletizing task from any position behind the safety barrier, and were encouraged to walk around the barrier in order to get a better understanding of the various robot's arm motions. Subjects from the *accident group* were exposed to the simulated accident immediately after this preview session.

During the experimental session, the subjects were asked to observe the robot for a 1-min period. After the robot traversed the path and its arm folded down in front of it to its resting position, the subject was instructed to approach the robot along one of the six angles preselected randomly, and stop before he or she could enter the robot's work envelope. The distance between the subject and the robot's work envelope was measured and recorded. The point of measurement on the subject was the tip of the toe. The subject was then requested to return to his or her position behind the safety barrier, and the experiment was repeated for all six randomly assigned angles of approach.

## 4. RESULTS

### 4–1. General Results and Discussion

While estimating the robot's working space, subjects crossed the true work envelope of the robots 49.6% of the time. The ANOVA table and main effects for analysis of the number of instrusions into the robot's work envelope are shown in Tables 8–2 and 8–3. With the exception of one interaction, the effects of all main variables and their interactions on the number of intrusions made were highly significant.

As shown in Tables 8–2 and 8–3, the effect of instruction type on the number of unintended worker intrusions into robot's true work envelope (perception of *reach zone* of the robot's arm versus *safe zone* around the robot) was highly significant ($p < 0.00001$). Workers invaded the robot's work envelope more

**TABLE 8–2 ANOVA Table for the Number of Intrusions into the Robot's True Work Envelope**

| Source | Degrees of freedom (d.f.) | Type I SS | $F$ | $p$ |
|---|---|---|---|---|
| Instruction type (IT) | 1 | 22.9 | 230.1 | 0.0001 |
| Accident exposure (AE) | 1 | 8.26 | 82.8 | 0.0001 |
| Robot size (RS) | 1 | 5.25 | 52.6 | 0.0001 |
| IT×AE | 1 | 4.16 | 41.8 | 0.0001 |
| IT×RS | 1 | 1.66 | 16.7 | 0.0001 |
| AE×RS | 1 | 0.08 | 0.85 | 0.3562 |
| IT×AE×RS | 1 | 0.91 | 9.2 | 0.0025 |
| Robot speed (robot size) | 2 | 2.06 | 10.3 | 0.0001 |
| Angle (robot size, speed) | 20 | 7.64 | 3.8 | 0.0001 |
| Error | 546 | 54.5 | | |
| Corrected total | 575 | 107.5 | | |

**TABLE 8–3 Main Effects for the Number of Intrusions into Robot's Work Envelope**

| Variable | N | Number of intrusions[a] | Percent of intrusions | Percent of total intrusions |
|---|---|---|---|---|
| Instruction type | | | | |
| Reach zone | 288 | a 129 | 44.7 | 90.21 |
| Safety zone | 288 | b 14 | 4.9 | 9.79 |
| | Total | 143 | 49.6 | 100.0 |
| Accident exposure | | | | |
| Accident group | 288 | a 37 | 12.9 | 25.87 |
| No-accident group | 288 | b 106 | 36.8 | 74.13 |
| | Total | 143 | 49.7 | 100.0 |
| Robot size | | | | |
| P50 | 288 | a 44 | 15.3 | 30.77 |
| MH33 | 288 | b 99 | 34.4 | 69.23 |
| | Total | 143 | 49.6 | 100.0 |
| Robot speed (cm/sec) | | | | |
| 25 | 288 | a 87 | 30.3 | 60.84 |
| 90 | 288 | b 56 | 19.4 | 39.16 |
| | Total | 143 | 49.7 | 100.0 |
| Angle of approach | | | | |
| Angle 1 | 96 | b 15 | 15.6 | 10.49 |
| Angle 2 | 96 | b 19 | 19.8 | 13.29 |
| Angle 3 | 96 | a 39 | 40.6 | 27.27 |
| Angle 4 | 96 | a 36 | 37.5 | 25.17 |
| Angle 5 | 96 | b 18 | 18.8 | 12.59 |
| Angle 6 | 96 | b 16 | 16.7 | 11.19 |
| | Total | 143 | 49.7 | 100.0 |

[a]Different letters indicate differences between means at $p \leq 0.05$.

often when instructed to determine the maximum reach of the robot's arm, than when asked to determine the minimum distance from the robot they perceived as safe. The above results indicate that when emphasis is placed on technological aspects, i.e., on the operational characteristics of robot's functioning, rather than on safety aspects, workers may come closer to the robot, and, therefore, may put themselves at a greater risk of injury.

An exposure to simulated industrial accident, robot size, robot speed, and angle of approach toward the robot significantly affected ($p < 0.0001$) distances chosen by the subjects. In general, workers from the *accident* group had fewer intrusions for both robots than those subjects who did not witness the accident. Compared to the frequency of intrusions with the robot speed of 25 cm/sec, the high speed of robot motions (90 cm/sec) kept workers away from the robot's envelope on a greater number of occasions.

Further analysis showed that 90.2% of all intrusion cases occurred under *reach zone* instructions, i.e., when workers attempted to perceive the maximum reach of the robot's arm, while only 9.8% of intrusions were attributed to the *safe zone* condition. Also, intrusions made by the workers who were exposed to the simulated accident accounted only for 25.9% of all intrusions, while intrusions made by the *no-accident* group accounted for 74.1% of the total.

The number of intrusions was significantly higher when working with the larger (MH33) robot (69.2% of total intrusions), than when working with the smaller robot (P50) (30.8% of total intrusions). More intrusions into robot's work envelope also occurred when the speed of robot motions was 25 cm/sec (60.8% of total intrusions) versus 90 cm/sec (39.2% of total intrusions).

**TABLE 8–4 Distribution of Intrusions into Robot's Work Envelope for Interactions among Instruction Type, Accident Exposure, and Robot Size**

| Variable | *N* | Number of intrusions[a] | Percent of intrusions | Percent of total intrusions |
|---|---|---|---|---|
| Reach zone | | | | |
| Accident group | 144 | a 35 | 24.3 | 27.13 |
| No-accident group | 144 | b 94 | 65.3 | 72.87 |
| | Total | 129 | 44.8 | 100.0 |
| Safe zone | | | | |
| Accident group | 144 | a 2 | 1.4 | 14.29 |
| No-accident group | 144 | b 12 | 8.3 | 85.71 |
| | Total | 14 | 4.9 | 100.0 |
| Reach zone | | | | |
| P50 | 144 | a 43 | 29.9 | 33.33 |
| MH33 | 144 | b 86 | 59.7 | 66.67 |
| | Total | 129 | 44.8 | 100.0 |
| Safe zone | | | | |
| P50 | 144 | a 1 | 0.7 | 7.14 |
| MH33 | 144 | b 13 | 9.0 | 92.86 |
| | Total | 14 | 4.9 | 100.0 |

[a]Different letters indicate differences between means at $p \leq 0.05$.

**TABLE 8–5 Distributions of Intrusions into Robot's Work Envelope by Robot Size and Speed of Robot Motions**

| Variable | Speed (cm/sec) | $N^a$ | Number of intrusions | Percent of intrusions | Percent of total intrusions |
|---|---|---|---|---|---|
| Reach zone | | | | | |
| Accident group | | | | | |
| P50 | 25 | 36 | 2 | 5.6 | 5.71 |
| | 90 | 36 | 1 | 2.8 | 2.86 |
| MH33 | 25 | 36 | 20 | 55.6 | 57.14 |
| | 90 | 36 | 12 | 33.3 | 34.29 |
| | Total | 144 | 35 | 24.3 | 100.00 |
| No-accident group | | | | | |
| P50 | 25 | 36 | 24 | 66.7 | 25.52 |
| | 90 | 36 | 16 | 44.4 | 17.02 |
| MH33 | 25 | 36 | 29 | 80.6 | 30.85 |
| | 90 | 36 | 25 | 69.4 | 36.60 |
| | Total | 144 | 94 | 65.2 | 100.00 |
| Safe zone | | | | | |
| Accident group | | | | | |
| P50 | 25 | 36 | 0 | 0 | 0 |
| | 90 | 36 | 0 | 0 | 0 |
| MH33 | 25 | 36 | 2 | 5.6 | 100.00 |
| | 90 | 36 | 0 | 0 | 0 |
| | Total | 144 | 2 | 1.4 | 100.00 |
| No-accident group | | | | | |
| P50 | 25 | 36 | 0 | 0 | 0 |
| | 90 | 36 | 1 | 2.8 | 8.33 |
| MH33 | 25 | 36 | 10 | 27.8 | 83.34 |
| | 90 | 36 | 1 | 2.8 | 8.33 |
| | Total | 144 | 12 | 8.3 | 100.00 |

[a]Number of observations

Those workers from the *safe zone* group who were exposed to simulated industrial accident did not intrude at all into the P50 robot's work envelope, while the intrusion for the MH33 robot accounted for less than 3% of the total for this group (see Tables 8–4 through 8–6). Finally, there were significantly more intrusions into the true work envelope when approaching the robot in its middle section, i.e., the two front paths (52.4% of total intrusions), than on the left two

**TABLE 8–6 Distribution of Intrusions into Robot's Work Envelope by Approach Angle for Each Robot**

| Variable | Angle of approach | *N* | Number of Intrusions | | Percent of Intrusions | | Percent of Total Intrusions | |
|---|---|---|---|---|---|---|---|---|
| | | | P50 | MH33 | P50 | MH33 | P50 | MH33 |
| Reach zone | | | | | | | | |
| Accident group | 1 | 12 | 0 | 2 | 0 | 16.7 | 0 | 6.25 |
| | 2 | 12 | 0 | 4 | 0 | 33.3 | 0 | 12.50 |
| | 3 | 12 | 3 | 10 | 25.0 | 83.3 | 100.00 | 31.25 |
| | 4 | 12 | 0 | 10 | 0 | 83.3 | 0 | 31.25 |
| | 5 | 12 | 0 | 3 | 0 | 25.0 | 0 | 9.375 |
| | 6 | 12 | 0 | 3 | 0 | 25.0 | 0 | 9.375 |
| | Total | 72 | 3 | 26 | 4.2 | 36.1 | 100.00 | 100.00 |
| No-accident group | 1 | 12 | 5 | 8 | 41.7 | 66.7 | 12.5 | 14.81 |
| | 2 | 12 | 5 | 8 | 41.7 | 66.7 | 12.5 | 14.81 |
| | 3 | 12 | 11 | 11 | 91.7 | 91.7 | 27.5 | 20.38 |
| | 4 | 12 | 9 | 11 | 75.0 | 91.7 | 22.5 | 20.38 |
| | 5 | 12 | 6 | 8 | 50.0 | 66.7 | 15.0 | 14.81 |
| | 6 | 12 | 4 | 8 | 33.3 | 66.7 | 10.00 | 14.81 |
| | Total | 72 | 40 | 54 | 55.6 | 75.0 | 100.00 | 100.00 |
| Safe zone | | | | | | | | |
| Accident group | 1 | 12 | 0 | 0 | 0 | 0 | 0 | 0 |
| | 2 | 12 | 0 | 0 | 0 | 0 | 0 | 0 |
| | 3 | 12 | 0 | 0 | 0 | 0 | 0 | 0 |
| | 4 | 12 | 0 | 1 | 0 | 8.3 | 0 | 50.00 |
| | 5 | 12 | 0 | 0 | 0 | 0 | 0 | 0 |
| | 6 | 12 | 0 | 1 | 0 | 8.3 | 0 | 50.00 |
| | Total | 72 | 0 | 2 | 0 | 2.8 | 0 | 100.00 |
| No-accident group | 1 | 12 | 0 | 0 | 0 | 0 | 0 | 0 |
| | 2 | 12 | 0 | 2 | 0 | 16.7 | 0 | 18.18 |
| | 3 | 12 | 0 | 4 | 0 | 33.3 | 0 | 36.36 |
| | 4 | 12 | 1 | 4 | 8.3 | 33.3 | 100.00 | 36.36 |
| | 5 | 12 | 0 | 1 | 0 | 8.3 | 0 | 9.10 |
| | 6 | 12 | 0 | 0 | 0 | 0 | 0 | 0 |
| | Total | 72 | 1 | 11 | 1.4 | 15.9 | 100.00 | 100.00 |

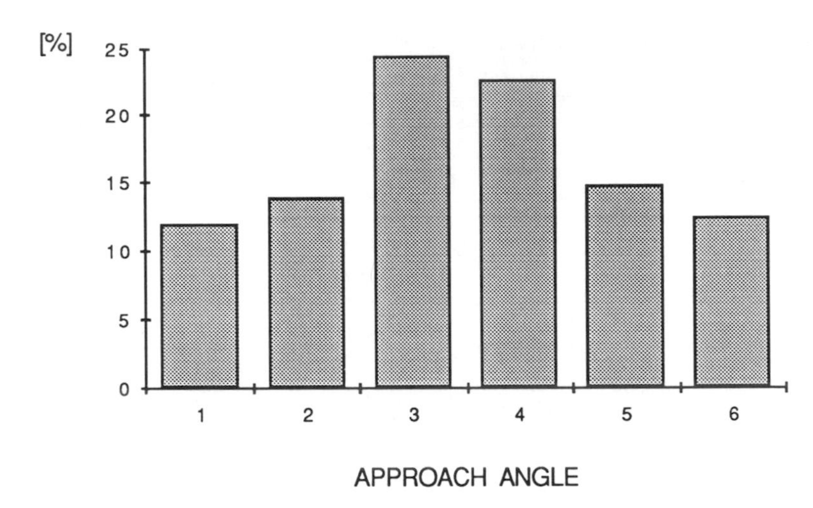


**FIGURE 8–2.** Percentage distribution of intrusions into robot's work envelope.

paths (23.7% of total intrusions) and the right two paths (23.7% of total intrusions). This effect of approach angle on the number of intrusions is illustrated in Figure 8–2.

## 5. CONCLUSIONS

The results of this study revealed that industrial workers frequently misperceive geometry of the true robot's work envelope, and may, therefore, frequently invade the robot's working space, especially when workplace layout places them in the middle section (front) of the robot. Clearly, more experimental research is needed to address other important aspects of worker perception of hazardous robotic workstations (see Karwowski et al., 1987; Karwowski and Rahimi, 1991), and to enhance our understanding of human–robot interactions, before effective safety guidelines for working with industrial robots can de developed.

## ACKNOWLEDGMENTS

The authors wish to acknowledge our technical staff, Mr. John Jones and Mr. Gary Graf, from the Factory Automation Laboratory, University of Louisville, for providing logistic support for this project. We are also indebted to Mrs. Laura Abell, Secretary, Center for Industrial Ergonomics, for her work on preparation of the manuscript. This project was supported by a grant from the National Institute of Occupational Safety and Health, No. 1 R01 OH02568-01.

**References**

Argote, L., Goodman, P. S., and Schkade, D. 1982. *The Human Side of Robotics: How Workers React to a Robot,* Technical Report 38-8182, Graduate School of Industrial Administration, Carnegie-Mellon University, Pittsburgh, PA.

Ayres, R. U. and Miller, S. M. 1983. *Robotics: Applications & Social Implications,* Cambridge, MA: Ballinger.

Backstrom, T. and Harms-Ringdahl, L. 1983. A statistical study on control system and accidents at work. In *Proceedings of the International Seminar on Occupational Accidents*, Stockholm, Sweden.

Carlsson, J. 1980. *Industrial Robots and Accidents at Work.* T RITA-ADG-0004, Stockholm, Sweden.

Carlsson, J. 1984. *Robot Accidents in Sweden.* National Board of Occupational Safety and Health, 5-171-84, Solna, Sweden.

Jiang, B. C. and Gainer, C. A., Jr. 1987. A case and effect analysis of robot accidents. *Journal of Occupational Accidents* 9:27–45.

Karwowski, W., Plank, T., Parsaei, M., and Rahimi, M. 1987. Human perception of the maximum safe speed of robot motions. *Proceedings of the Human Factors Society-31st Annual Meeting.* pp. 186–190. Human Factors Society, Santa Monica, CA.

Karwowski, W. and Rahimi, M. 1991. Worker selection of safe speed and idle condition in simulated monitoring of two industrial robots. *Ergonomics* 34(5):531–546.

Linger, M. 1988. *Robot Safety—Are Robots Safe?* Göteborg, Sweden: Swedish Institute of Production Engineering Research.

Mastersson, L. 1987. Interaction between man and robots with some emphasis on "intelligent" robots. In *Occupational Health and Safety in Automation and Robotics,* ed. K. Noro. pp. 143–150. UK: Taylor & Francis.

Nagamachi, M. 1986. Human factor of industrial robots and robot safety management in Japan. *Applied Ergonomics* 17(1):9–18.

Nagamachi, M. 1988. Ten fatal accidents due to robots in Japan. In *Ergonomics of Hybrid Automated Systems I.* eds. W. Karwowski et al. pp. 391–396. Amsterdam: Elsevier.

Noro, K. and Okada, Y. 1983. Robotization and human factors. *Ergonomics* 26:985–1000.

Parsons, H. M. 1985. Human factors in industrial robot safety. *Journal of Occupational Accidents* 8:25–47.

Parsons, H. M., and Kearsley, G. P. 1982. Robotics and human factors: current status and future prospects. *Human Factors* 24:534–552.

Rahimi, M., and Hancock, P. A. 1986. Perception-decision-and-action processes in operator collision avoidance with robots. *Proceedings of the Annual Conference of the Human Factors Association of Canada.*

Sanderson, L. M., Collins, J. N., and McGlothlin, J. D. 1986. Robot-related fatality involving a U.S. manufacturing plant employee: case report and recommendations, *Journal of Occupational Accidents.* 8:13–23.

Sheridan, T. G. 1984. Supervisory control of remote manipulators, vehicles and dynamic processes: experiments in common and display aiding. *Advances in Man-Machine Systems Research*. ed. N. B. Rouse Vol. 1. Greenwich: JAI Press.

Sugimoto, N. 1985. Safety measures for automated machines. *Safety Staff* 12:4–18.

Sugimoto, N. and Kawaguchi, K. 1983. Fault tree analysis of hazards created by robots. *Robot Safety,* eds. M. C. Bonney and Y. F. Yong, pp. 83–98. Berlin: Springer-Verlag, IFS Ltd.

Yokomizo, Y., Hasegawa, Y., and Komatsubara, A. 1987. Problems of and industrial medicine measures for the introduction of robots. *Occupational Health and Safety in Automation and Robotics,* ed. K. Noro. pp. 167–174. London: Taylor and Francis.

# 9

# Reliability-Based Control for Intelligent Machines

John E. McInroy and George N. Saridis

## 1. INTRODUCTION

As the study of robotic systems has progressed throughout the years, a large archive of special-purpose algorithms suitable for various control and sensing tasks has been produced. Often several different algorithms are available that are capable of executing the same task with varying levels of performance. Use of redundant measurements, for example, improves accuracy at the cost of increased computational time. Similarly, a vision system may be capable of producing depth measurements using feature matching, point matching, focusing, structured lighting, stadimetry, etc. Control strategies may include either a PD or PID compensator for the same task, thus yielding different accuracies and response times. Classically designed compensators may be well suited for meeting a desired settling time and overshoot, while optimally designed compensators are best at meeting a quadratic cost functional.

In light of the plethora of algorithms available for treating many common robotic tasks, this work does not derive new control/ sensing algorithms, but rather presents a consistent and general framework for analyzing the reliability of these algorithms. As such, it contributes to the final goal of designing "intelligent machines" capable of operating in uncertain environments with minimal supervision or interaction with a human operator. Other aspects of this subject have already been extensively developed by Saridis, who has proposed the combination of artificial intelligence, control systems, and operations research through the use of information theory (Saridis, 1988; Saridis and Valavanis, 1988; Valavanis and Saridis, 1988).

Since many of the available algorithms overlap in application, direct comparisons can often be made. This has produced a growing trend to assess the accuracy of particular algorithms under consideration. Unfortunately, increased precision is often gained at the cost of increased time, computations, complexity, etc. Consequently, a sensing algorithm often cannot be fully evaluated in itself, but must be considered in conjunction with the control system that produces actions based on the measured perceptions. Similarly, the control performance depends on the sensing statistics. In short, control and sensing are dual entities: control determines interaction with the environment, while sensing observes the effect of these interactions.

The coordination of control and sensing in a mathematically rigorous fashion is a generalization and extension of sensor fusion research termed *reliable control and sensor fusion*. Reliable control and sensor fusion is defined as the unification of sensor information and control strategies such that an acceptable level of reliability in accomplishing a desired task is achieved. Techniques to fulfill this goal based on entropy measures are explained in McInroy and Saridis (1990a). This work brings together all of these concepts so that the reliability analysis can begin with rapid generation of entropy constraints from the specifications, and end with the reliability self-information terms corresponding to each feasible plan. The statistical methods for estimating reliability are refined, with maximum likelihood estimation used when applicable, and a lower bound on the reliability presented. In addition, recent advances in vision analysis (Lee and Kay, 1990; Rodriguez and Aggarwal, 1990) are incorporated into the framework. These advances, combined with the position analysis contained in McInroy and Saridis (1990b), allow realistic evaluation of high-precision hand/eye coordination, as will be demonstrated in the case study.

Previous research on individual issues has contributed greatly to certain segments of this work, but very few attempts have been made to bring these topics together. Most reliability studies to date have focused on the performance of a particular algorithm. Azadivar (1987) finds the reliability of positioning a tool due to random joint positioning errors. Blostein and Huang (1987), as well as Rodriguez and Aggarwal (1990), determine the accuracy of stereo vision position measurement contaminated with pixel truncation errors. Lee and Kay (1990) complement this analysis by analyzing the accuracy of a stereo vision system in measuring both position and orientation of an object. Considerable work has been aimed at combining alternative sensing strategies without regard to control. Smith and Cheeseman (1986) integrate position measurements of a mobile robot from several homogeneous coordinate frames by propagating covariance matrices. Durrant-Whyte (1988) uses a similar approach in propagating vectors of geometrical parameters between coordinate frames.

The few attempts at assessing the reliability of entire plans have not approached the problem from a completely probabilistic viewpoint. Havel and Kramosil (1978) propose reliability for use in robot plan formation, but assign reliability using heuristic techniques. Smith and Gini (1986) use "the experience, intuition, and common sense of the robot programmer" to anticipate unreliable plans. Brooks (1982) develops a "plan checker" that evaluates the feasibility of a plan, performing the analysis based on a geometrical analysis of tolerances. Taylor and Taylor (1988) and Taylor et al. (1990) utilize a probabilistic framework in error identification and recovery.

This work suggests a method for unifying reliability, control, and sensing. Moreover, rather than relying on heuristic simplifications or highly constrained tasks, a general, mathematical framework founded on entropy is derived. First, it is shown that reliability specifications can be mapped to a set of entropy constraints that define the precision necessary for the task at hand. Next, an algorithm is presented for reliable control and sensor fusion that utilizes the specification entropy. Entropy is shown to be invariant with respect to homogeneous coordinate frame transformations, thus simplifying the reliability analysis. By using reliability and information theory, techniques are developed to provide an analytical means of evaluating the reliability of a sequence of elementary events. Moreover, by augmenting reliability theory with information-theoretic concepts, simple methods of combining the reliabilities of different coordinator subsystems are obtained. To validate these concepts, a detailed analysis of an extremely common robotic task is presented—the problem of vision-guided positioning to an oriented point.

## 2. SELECTION OF FEASIBLE PLANS

The Organizer (or a human teleoperator, in the case of telerobotics) in an Intelligent Control System formulates the most probable complete and compatible plan that will execute the user-requested job in the best possible way. The reliability analysis aids this process by evaluating alternative plans to assess their feasibility. To summarize, reliability can be used to solve the following problem:

> Given an explicit task to be executed (as provided by the Organizer), first separate from the entire space of plans $A = \{A_1, \ldots, A_n\}$ those subsets of algorithms that are potentially capable of attaining performance within the desired specifications (feasible plans). For these feasible plans ($A_{\text{feas}_i}$), find the reliabilities associated with the alternative subsets of control and sensing algorithms such that the task can be accomplished to meet the set of desired specifications, $S_D = \{s_1, \ldots, s_m\}$.

This section focuses on the first step, the selection of feasible algorithms. Since this work is an analysis technique, it does not consider the more intelligent activities required to formulate the sets of algorithms. Rather, it explores the statistical behavior of alternative plans. Thus [in the context of Intelligent Machines as proposed by Valavanis and Saridis (1988)] the Organizer and

Coordinators first select, from the library of all possible control and sensing algorithms, subsets of algorithms potentially able to solve the given task. These subsets may well contain multiple control algorithms for applications such as gross positioning followed by fine movement. Moreover, multiple-sensing algorithms may be included for utilizing several sensor subsystems. In fact, even redundant measurements may be used for noise suppression and fault tolerance. A particular subset is denoted as $A_i$, and may be regarded as a low-level plan for executing the desired task. On the other hand, not all algorithms are included because some obviously do not apply. For instance, if position measurement is required, vision routines for object inventory are not included. For the most part, this portion of the planning utilities logical predicates that are adroitly handled through the use of Petri Net Transducers (Wang and Saridis 1989). In contrast, once these plans are formulated, it is still necessary to examine the statistical characteristics of the plans, $A_i$, and their relationship to the set of desired specifications, $S_D$.

More explicitly, the present goal is separating from the entire space of plans $A = \{A_1, \ldots, A_n\}$ those subsets of algorithms that are capable of attaining performance within the desired specifications.

In order to readily incorporate the design specifications into the reliability analysis, several concepts from information theory are employed.

## 2–1. An Entropy Approach to Specifications

In the context of Intelligent Machines, *reliability* can be defined as the probability of success in executing a task. The task is termed a *success* if execution of the task is accomplished such that a desired set of specifications, $S_D$, is satisfied. In other words, it has long been recognized that perfect positioning, measurement, etc., is not possible, so a "window" of acceptable range is embodied in the specifications. This concept is formalized in physics by Heizenberg's Uncertainty Principle, although engineers have included specifications as part of a design throughout recorded history.

In order to define the interaction of the specifications with control and sensing strategies, it is first necessary to express the specifications in an information-theoretic setting using Jaynes maximum entropy method (MEM).

Jaynes maximum entropy method is a procedure for determining the least-biased probability distribution of a probability space subject to given constraints. The least-biased distribution is that distribution that maximizes the entropy subject to the given constraints (Jaynes 1957, 1982). The MEM has been applied to many statistical inference questions, and several classes of problems have useful and tractable solutions. For instance, suppose we wish to find the density, $f(x)$, of a random variable $x$ subject to the condition that the expected values, $\mu_i$, of $n$ known functions $g_i(x)$ are given. MEM analysis yields a density

$$f(x) = A \exp[-\lambda_1 g_1(x) - \cdots - \lambda_n g_n(x)], \tag{9–1}$$

where A, $\lambda_1$, ..., $\lambda_n$ are constants derived from the preassigned constraints (Papoulis, 1984). The maximum entropy corresponding to the constraints is

$$H(x) = \lambda_1\mu_1 + \cdots + \lambda_n\mu_n - \ln A. \tag{9–2}$$

Consequently, if constraints involving expected values (i.e., moments, etc.) are known, then the MEM, through (9–1) and (9–2), provides the least-biased distribution and entropy. MEM distributions and entropies for several common sets of constraints are presented in Harr (1987) and Kalata and Priemer (1974). For convenience, Table 9–1 summarizes those results.

The key to unifying the reliability and information-theoretic approaches lies in the MEM of Jaynes. Given the set of desired specifications, $S_D$, it is possible using the MEM to generate the least-biased probability density function given the specifications. Moreover, the maximum (worst case) entropy corresponding to the given specifications is also provided by the method. For instance, if the specification is a safe velocity range, $(-v_{max}, v_{max})$, then the MEM yields a uniform distribution as the maximum entropy distribution, with $H$(specification) $= \ln 2v_{max}$ (Table 9–1). Similarly, if the specification is a positioning error with mean of zero and variance $\sigma^2_{e_p}$, then the maximum entropy distribution is Gaussian with $H$(specification) $= \ln \sqrt{2\pi e\sigma^2_{e_p}}$.

**TABLE 9–1 Maximum Entropy Distributions and Entropies for Several Common Sets of Constraints**

| Constraints | Distribution | Density | Entropy |
|---|---|---|---|
| Bounds, $(a,b)$ | Uniform | $\dfrac{1}{b-a}$ | $\ln[b-a]$ |
| Mean, $\mu$ | Exponential | $\dfrac{1}{\mu} e^{-x/\mu}$ | $\ln[\mu e]$ |
| Mean $(\mu)$, Variance $(\sigma^2)$ | Gaussian | $\dfrac{1}{\sqrt{2\pi\sigma^2}} \exp\left[-\dfrac{(x-\mu)^2}{2\sigma^2}\right]$ | $\dfrac{1}{2}\ln[2\pi e\sigma^2]$ |
| Bounds (a,b), Mean $\mu$ | Truncated exponential | $\dfrac{\lambda}{e^{-\lambda a} - e^{-\lambda b}} e^{-\lambda x}$, | $\lambda\mu$ |
| | | $\mu = \dfrac{1}{\lambda} + \dfrac{be^{-\lambda b} - ae^{-\lambda a}}{e^{-\lambda a} - e^{-\lambda b}}$ | $-\ln\left[\dfrac{\lambda}{e^{-\lambda a} - e^{-\lambda b}}\right]$ |

In effect, the set of specifications ($S_D$) include into the design an allowable level of uncertainty. The MEM facilitates the expression of this design uncertainty as a Shannon entropy. Once formulated in this manner, several important concepts from information theory may be invoked. For instance, information theorists have long noted that entropy is analogous to information. Consequently, the specifications needed for reliable operation can, through MEM analysis, be incorporated into information flows within an Intelligent Machine. Moreover, Saridis and Valavanis (1988) define *knowledge* as a form of structured information. As a result, the knowledge embodied in the design specifications can be mathematically represented through the information-theoretic approach. A method for utilizing this knowledge as a criterion for reliable control and sensor fusion is presented in the next section.

## 2–2. Generation of Entropy Constraints

The entropy formulation of the specifications allows well-developed concepts from information theory and intelligent control theory to be combined and utilized in the plan selection. In this context, a plan consists of a subset of control and sensing algorithms, $A_i = \{C_i,S_i\}$, where $C_i$ and $S_i$ are control and sensing strategies for the $i$th plan, respectively. Thus, by propagating the entropies of distributed sensor readings to the control coordinate frame and fusing these entropies with that of the controller, the total entropy $H$(control, sensing) can be compared to the maximum entropy allowed by the specifications, $H$(specification). If $H$(control, sensing) exceeds $H$(specification), then that set of sensing and control is not capable of meeting the specification, and will therefore not be included in the feasible algorithms, $A_{feas} = \{A_{feas_1}, A_{feas_2}, \ldots, A_{feas_j}\}$. This induces a formal definition for feasible plans:

**Definition:** Given an explicit task to be executed and a subset of control and sensing algorithms, $A_i = \{C_i,S_i\}$, corresponding to that task, $A_i$ is a *feasible plan* (denoted by $A_{feas_i}$) if the entropy constraints

$$H(\mathrm{A}_{i_k}) \leq H(s_k) \tag{9–3}$$

are satisfied for all of the specifications, $s_k$, $k = 1, \ldots, m$. $H(A_{i_k})$ is the entropy of $A_{feas_i}$ in responding to the $k$th specification, and $H(s_k)$ is the uncertainty embodied in the $k$th specification.

This definition yields a set of entropy constraints that must be satisfied to ensure reliable operation. Thus, once all entropies have been determined, finding those plans that are feasible can be accomplished in a straightforward manner from the entropy constraints as depicted in Figure 9–1.

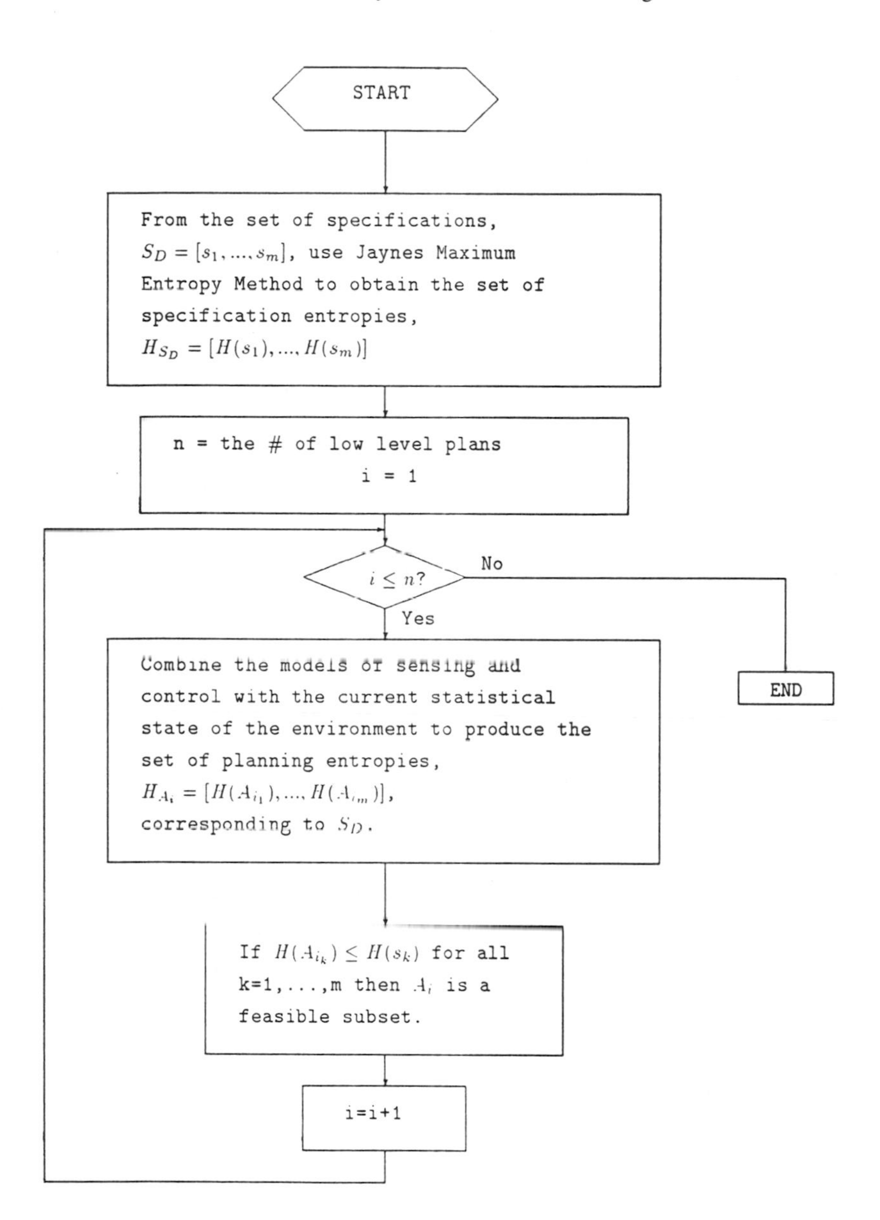


**FIGURE 9–1.** A flowchart for selecting feasible plans.

Intuitively, Eq. (9–3) implies that the uncertainty regarding a feasible subset's ability to meet each design criterion must be less than the permissible level of uncertainty embodied in each design specification. To be more concrete, consider what is perhaps the most common design scenario—a specification consisting of a tolerance range $(a,b)$ and a feasible plan whose response is $A \sim N([a + b]/2, \sigma^2)$. This may represent a positioning tolerance, a desired force and safe bounds, velocity constraints, etc. For feasibility, (9–3) and Table 9–1 imply that

$$\ln \sqrt{2\pi e \sigma^2} \leq \ln[b - a], \tag{9–4}$$

or, equivalently,

$$4.13\sigma \leq b - a. \tag{9–5}$$

For this problem, the reliability is the probability that $A$ stays within the tolerance range, i.e.,

$$R = \Pr\{a \leq A \leq b\}. \tag{9–6}$$

Since the response distribution is Gaussian, the reliability is easily found:

$$R = 2\Phi\left(\frac{b - a}{2\sigma}\right) - 1 \tag{9–7}$$

where $\Phi(\cdot)$ is the standard normal cumulative distribution function. The worst case reliability allowed by (9–5) is $4.13\sigma = b - a$, which produces $R = 0.96$. Thus, the entropy constraints (9–3) yield a 0.96 reliability for Gaussian responses subject to tolerance specifications. Consequently, the implicit reliability contained in the entropy constraints is sufficient to ensure satisfactory performance for this example.

Note that, in this scenario, the mean of the plan and that of the specification are identical. Moreover, both distributions are symmetrical about their means. This symmetry ensures that the plan uncertainty and the specification uncertainty pertain to the same physical parameters; without this property, the entropy constraints produce less reliable results. Consequently, it is important to strive for this symmetry when formulating the problem, as will be further illustrated in the case study (Section 4).

Reliable control and sensor fusion possess several advantages due to their information-theoretic roots. These advantages will be discussed in the next section.

## 2–3. Advantages of the Information-Theoretic Approach

The information-theoretic approach to reliable control and sensor fusion holds several advantages over other methods of analysis. These advantages can be divided into three categories. First, the entropy associated with many sensed features useful in robotics is invariant with respect to coordinate frame transformations. Second, since the MEM is a well-developed procedure of statistical inference, it provides exact methods of handling uncertainty for any distribution. Third, because entropy can be interpreted as information, it provides a consistent representation throughout all levels in a hierarchically Intelligent Machine.

First, consider entropy in the presence of homogeneous coordinate transformations. The joint entropy of many diverse types of measurements is invariant with respect to coordinate frame transformations (Kyriakopoulos 1988). This is easily shown by noting that the joint entropy of a random vector $x_j = g(x_i)$ is found by (Papoulis, 1984):

$$H(x_j) = H(x_i) + E\{\ln |\det(J)|\}, \qquad (9\text{–}8)$$

where $J$ is the Jacobian of the uniquely invertible transformation $g(\cdot)$, $J = \nabla_x g(x)$. For coordinate transformations, $x_j = g(x_i) = T_i^j x_i$ (Fu et al., 1987), where $x_j$ is a point represented in the $j$th coordinate frame, $x_i$ is a point represented in the $i$th frame, and

$$T_i^j = \begin{bmatrix} n & o & a & q \\ 0 & 0 & 0 & 1 \end{bmatrix} = \begin{bmatrix} R & q \\ 0 & 1 \end{bmatrix}, \qquad (9\text{–}9)$$

where $n$, $o$, $a$, and $q$ are vectors, and $R$ is an orthonormal rotation matrix. The Jacobian of this transformation is simply $J = T_i^j$. Since $R$ is orthonormal, $\det(J) = \det(R) * 1 = 1$. Substitution into (9–8) then yields $H(x_j) = H(x_i)$. This fact was first noted by Kyriakopoulos (1988). Kyriakopoulos also found that the result could be generalized far beyond mere 3-D points to include parameter vectors of several common geometric features: lines, planes, and spheres. In this formulation, a geometric feature such as a line is represented as a point (parameter vector) in parameter space following Durrant-Whyte (1988). The joint entropy of the entire parameter vector is then invariant with respect to coordinate frame transformations. A similar result pertains for six-dimensional quantities such as oriented points in $\mathbb{R}^3$ or force vectors. The transformations for several useful entropy invariant parameter vectors are listed in Table 9–2.

This invariance with respect to coordinate frame transformations is very important when comparing and combining sensor readings distributed over many different coordinate frames—a very common situation in advanced robotic sys-

**TABLE 9–2 Features that Display Coordinate Frame Entropy Invariance**

| Feature | Equation | Parameters | Transformation |
|---|---|---|---|
| Point | | $p = [x\ y\ z\ 1]^T$ | $p_j = T_i^j p_i$ |
| Line | $r(t) = r_0 + td$ | $p = [d^T\ 1\ r_0^T\ 1]^T$ | $p_j = \begin{bmatrix} T_i^j & 0 \\ 0 & T_i^j \end{bmatrix} p_i$ |
| Plane | $n^T r = 1$ | $p = [n^T\ 1]^T$ | $p_j = T_j^{i^T} p_i$ |
| Sphere | $\|\|r - r_0\|\| = d$ | $p = [r_0^T\ 1\ d]^T$ | $p_j = \begin{bmatrix} T_i^j & 0 \\ 0 & 1 \end{bmatrix} p_i$ |
| Circle | $[n^T r = 1] \cap \|\|r - r_0\|\| = d$ | $p = [n^T\ 1\ r_0^T\ 1\ d]^T$ | $p_j = \begin{bmatrix} T_j^{i^T} & 0 & 0 \\ 0 & T_i^j & 0 \\ 0 & 0 & 1 \end{bmatrix} p_i$ |
| Oriented point | | $p = [x\ y\ z\ \theta_x\ \theta_y\ \theta_z]^T$ | $p_j = \Gamma_i^j(p_i)$<br>$J = J_i^j$ |
| Force | | $p = [f_x f_y f_z f_{\theta_x} f_{\theta_y} f_{\theta_z}]^T$ | $p_j = J_j^{i^T} p_i$ |

tems. To illustrate, cameras are often mounted on gimbals or on a link of the manipulator. Similarly, mobile robots perceive the environment from constantly changing positions. Even if the sensor itself is immobile, the relationship of the sensor with respect to the end effector's coordinate frame will vary as the manipulator position changes. In addition, often several different sensors are distributed throughout the workcell. For instance, a laser range finder often complements a stereo vision system. The efficient fusion of such measurements is of great importance.

Since modern robotic systems contain many sensors distributed over a variety of different physical locations that may change with time, transforming these diverse sensor readings to a common coordinate frame has posed one of the fundamental issues of sensor fusion research. Recently, attention has been focused on transforming both the sensed measurement and the uncertainty regarding that measurement, as the uncertainty may be used in the fusion of sensor readings. Most authors have expressed this uncertainty through the covariance matrix (Durrant-Whyte, 1988; Smith and Cheeseman, 1986). A more compact and useful representation of the uncertainty is Shannon's entropy. Shannon's definition of entropy has gained wide acceptance because it complies with a heuristic understanding of uncertainty and provides a mathematical framework to express this uncertainty (Papoulis, 1984). Moreover, entropy can be used in estimation to replace covariance statistics (Kalata and Priemer, 1974, 1979).

In contrast, the covariance matrix varies under coordinate frame transformations. Propagation of the covariance matrix can be approximately calculated as (Durrant-Whyte, 1988)

$$C_j = J_i^j C_i J_{iT}^j \tag{9–10}$$

where $C_j$ is the covariance in the $j$th frame. $J_i^j$ is the Jacobian matrix of the coordinate transformation, and $C_i$ is the covariance in the $i$th frame. As a result, in order to evaluate the performance of a distributed sensor in the control frame using covariance analysis, it is first necessary to calculate the Jacobian between those frames, and then find the propagated covariance using (9–10). In contrast, since the joint entropy is invariant with respect to coordinate frame transformations, it is not necessary to find the Jacobian when using entropy-based analysis. Because many different measurements are often available at a variety of coordinate frames, this simplification makes comparison between alternative sensing strategies much more computationally tractable. In all fairness, it must be stated that the complete covariance matrix is sometimes a useful quantity to be propagated in its own right. This is especially true when the desired specifications concern individual elements of the feature's parameter vector. However, even in these cases the information-theoretic approach still holds several important advantages.

To be explicit, the information-theoretic approach is advantageous because it is a well-developed statistical inference technique. As a result, the wealth of previous knowledge contained in information-theory literature may be immediately drawn upon. For instance, the entropy formulation expresses *uncertainty* in a meaningful fashion for any probability distribution. In contrast, propagating merely the covariance is sufficient in expressing the uncertainty only for certain distributions such as those listed in Kalata and Priemer (1979). Other distributions require higher-order moments for complete analysis of the uncertainty.

The final advantage that the unification of reliability and entropy concepts presents is the consistent representation of uncertainty throughout all levels of a hierarchically Intelligent Machine. Since entropy can be regarded as information, it is a sufficient analytic measure that unifies the treatment of all the levels of an intelligent machine (Saridis and Valavanis, 1988). This implies, for instance, that, in an impasse, the higher-level Organizer may redesign the problem, thus generating new specifications for the lower-level reliable control and sensor fusion. An iterative design procedure is then possible in the spirit of Koomen (1985). In addition, since entropies are produced for each feasible subset and each specification, the techniques have great potential for integration with other well-established results of Intelligent Control. For instance, if a positioning specification is phrased in terms of a maximum integral error, then it may be possible to make use of entropy formulations of optimal control (Saridis, 1988).

## 3. RELIABILITY ANALYSIS OF FEASIBLE PLANS

In order to synthesize control and sensing with reliability, a two-step analysis is used. First, the set of entropy constraints is found and is used to obtain feasible sets of algorithms (see Section 2). For many applications, this alone will yield sufficient reliability for the task at hand. This is especially true when failure imposes only a small penalty. On the other hand, when a high degree of reliability must be ensured, a second stage of analysis that explicitly calculates the reliability ($R_i$) corresponding to each feasible subset ($A_{\text{feas}_i}$) is necessary. Just as the specifications play an integral role in the determination of feasible plans, $S_D$ also greatly influences $R_i$. The set of specifications may include constraints on total execution time, positioning accuracy, maximum overshoot, robustness, tracking errors, etc. Each element of the specifications, $s_k$, represents a desired characteristic to be achieved during task execution, while $R_i$ incorporates all of these criteria into a single term and measures the probability that they will all be satisfied. This measurement is accomplished by first calculating the reliability of the plan in responding to each separate specification ($R_{ik}$, $k = 1, \ldots, m$). Next, the individual reliability terms are combined to form $R_i$.

### 3–1. Calculation of Reliability Terms

The individual reliability terms ($R_{ik}$) can be found by first defining reliability performance functions (RPF), $g_{ik}$, associated with each feasible subset, $A_{\text{feas}_i}$, and each specification, $s_k$. Often, this is the most arduous phase of the reliability analysis, since it requires capturing the probabilistic behavior of a particular algorithm and expressing it in terms of the desired specifications. The performance functions should be defined in such a manner that

$$[g_{ik}(x) > 0] \Rightarrow \text{success,} \tag{9–11}$$

$$[g_{ik}(x) \leq 0] \Rightarrow \text{failure,} \tag{9–12}$$

where $x$ is a vector of state variables.

Once each RPF is defined, the statistics associated with the current state of the environment can be used to find the reliability, $R_{ik}$, of the particular subset $A_{\text{feas}_i}$ in meeting the desired specification $s_k$. If the form of the underlying distribution for the state variables is known (based on prior experience), while the distribution parameters must be found from statistical sampling, then maximum likelihood estimation may be applied.

Maximum likelihood estimation is useful because under certain regularity conditions it possesses several compelling advantages. The regularity conditions are not very restrictive. In short, if samples $\{x_1, x_2, \ldots, x_k\}$ are taken from an underlying cumulative distribution function $P(x;e)$, where $e$ is a distribution parameter to be estimated, then the maximum likelihood estimate of $e$, $\hat{e}$, meets the regularity conditions if $P(x;e)$ is regular with respect to its first two $e$ derivatives (directional derivative) and $\hat{e}$ is unique. $P(x;e)$ is *regular with respect to its first e derivative* if

$$E\{S(x;e)\} = \frac{\partial}{\partial e}\int_{-\infty}^{\infty} dP(x;e) = 0, \tag{9–13}$$

where

$$S(x;e) = \frac{\partial}{\partial e} \log dP(x;e). \tag{9–14}$$

$P(x;e)$ is *regular with respect to its second e derivative* if

$$E\{S'(x;e)\} + E\{S(x;e)\}^2 = \frac{\partial^2}{\partial e^2}\int_{-\infty}^{\infty} dP(x;e) = 0, \tag{9–15}$$

where

$$S'(x;e) = \frac{\partial}{\partial e} S(x;e). \tag{9–16}$$

Consult Wilks (1962) or Zacks (1971) for a rigorous treatment of maximum likelihood estimation. Under these conditions, the maximum likelihood estimate (MLE) is asymptotically normal, consistent, and asymptotically efficient. Furthermore, for a function of $e$, $f(e)$, the MLE is given by the invariance property to be $f(\hat{e})$ (Zacks 1971). The distribution of the function is

$$\hat{f}(\hat{e}) \sim AN(f(\hat{e}), \nabla_e^T f(\hat{e})\mathrm{Cov}(\hat{e})\nabla_e f(\hat{e})), \tag{9–17}$$

where $\hat{f}(\hat{e})$ = the MLE of $\hat{f}(\hat{e})$
$AN$ = asymptotic normality
$\nabla_e f(\hat{e})$ = the gradient of $f(\hat{e})$ with respect to $\hat{e}$
$\mathrm{Cov}(\hat{e})$ = the covariance matrix of $\hat{e}$

Consequently, if the regularity conditions are satisfied for the underlying distribution, then from the sample ensemble $\{x_1, x_2, \ldots, x_k\}$ it is possible by maximizing the likelihood and using (9–17) to find an asymptotically normal maximum likelihood distribution for $g_{ik}(x)$. Since the distribution is asymptotically normal, it is a simple matter to then find the reliability corresponding to the RPF.

If maximum likelihood estimation cannot be used, a lower bound on $R_{ik}$ may be found from the reliability index. A lower bound is extremely useful, since it provides a guaranteed minimum level of reliability. The disadvantage, of course, is that the lower bound can be a conservative estimate of the actual reliability. The *reliability index,* $\beta$, is defined as the minimum distance between the origin of a set of uncorrelated standard normal variates (derived from $x$), and the failure surface, $g(x) = 0$.

Consider the case in which the individual state variables, denoted now as $x_i$, are uncorrelated, Gaussian random variables. The variables can be replaced by a set of reduced variates with the transformation:

$$x_i' = \frac{x_i - \mu_{x_i}}{\sigma_{x_i}}; \qquad i = 1,2, \ldots, n, \tag{9–18}$$

where $\mu_{x_i}$ is the expected value of the $i$th random variable, and $\sigma_{x_i}$ is the standard deviation of the $i$th random variable. To estimate the reliability, Shinozuka (1983) has shown that the point on the failure surface $[g(x) = 0]$ with the minimum distance to the origin of the reduced variates, $x'$, is the most probable failure point. If $g(x)$ is a nonlinear function, the reliability index may be used as an approximate measure of reliability by placing a lower bound on the reliability. For linear $g(x)$, the reliability can be found exactly from the reliability index.

Determining the reliability index, $\beta$, may require iterative methods for nonlinear performance functions. In contrast, linear performance functions have a closed-form solution. Suppose that the performance function is represented as

$$g(x) = a_0 + \sum_i a_i x_i, \tag{9–19}$$

where the $a_i$'s are constants. Note that the reliability performance function may be negative, although this possibility corresponds to a system failure. The minimum distance to the origin of the reduced variates is then (Ang and Tang, 1984)

$$\beta = \frac{a_0 + \Sigma_i a_i \mu_{x_i}}{\sqrt{\Sigma_i (a_i \sigma_{x_i})^2}} . \tag{9–20}$$

For linear performance functions, the reliability can be found directly from the reliability index by the formula

$$R = \Phi(\beta), \tag{9–21}$$

where $\Phi(\cdot)$ is the normalized, zero mean, Gaussian cumulative distribution function and $R$ is the reliability. The minimum distance for nonlinear RPFs can be found by iteratively searching for the minimum of

$$\beta = \frac{-\nabla g^T x'}{(\nabla g^T \nabla g)^{1/2}} \tag{9–22}$$

subject to

$$g(x') = 0 \tag{9–23}$$

where $\nabla g$ is the gradient vector of $g(x)$ with respect to $x'$, and $T$ denotes the transpose. Parkinson (1980) offers a method for efficient solution to the minimization.

If $g(x)$ is concave toward the origin of the reduced variates, then (Parkinson, 1982)

$$\chi_n^2(\beta^2) \leq R \leq \Phi(\beta). \tag{9–24}$$

On the other hand, if $g(x)$ is convex toward the origin, then

$$\Phi(\beta) \leq R \leq 1, \tag{9–25}$$

where $\chi_n^2(\cdot)$ is the chi-squared distribution function with $n$ degrees of freedom. If the RPF is neither convex nor concave, (9–24) may always be used as a lower bound on the reliability. Figure 9–2 graphically depicts the bounds for a two dimensional problem.

Should the variables be correlated, the solution is still possible. However, the covariance matrix must be used (Shinozuka, 1983). For non-Gaussian variates, Ang and Tang (1984) and Parkinson (1982) explain techniques for transforming the variates to an equivalent Gaussian system.

### 3–2. Combination of Reliability Terms

Once the reliabilities in meeting each individual specification have been found, the ability of the potential algorithms ($A_{\text{feas}_i}$) in meeting the desired specifications, $S_D$, depends on the relationships between the elements of $S_D$. In the

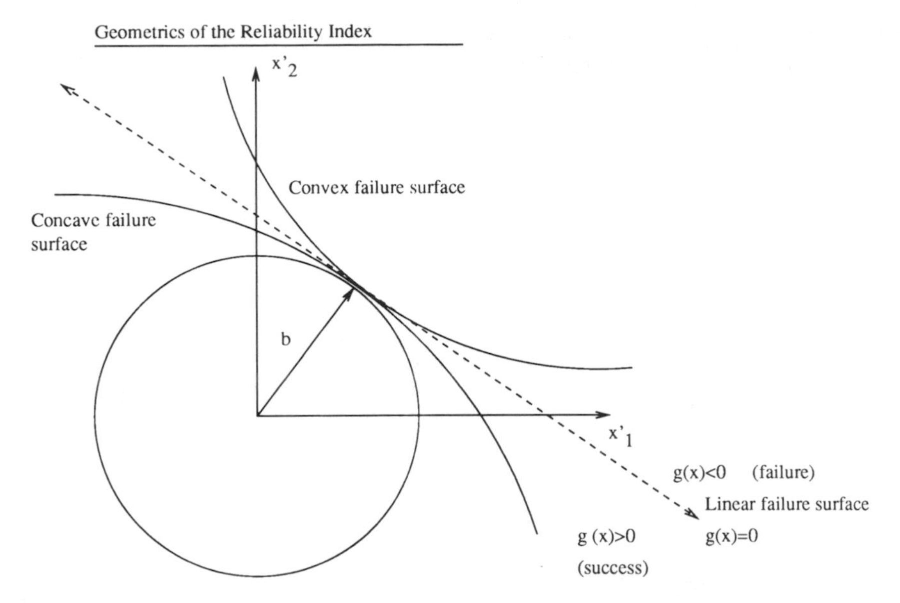


**FIGURE 9–2.** A geometrical picture of the reliability index for two dimensions.

majority of cases, the elements $s_k$ will have a series relationship, i.e., the success in meeting requirements will depend on all specifications being satisfied. Occasionally, the elements may have parallel relationships. For instance, it may be desired to either meet a desired overshoot *or* a desired execution time. Since meeting either criterion is termed a success, the relationship is parallel.

To allow ease in combining reliabilities, as well as to accommodate information flows within Intelligent Machines, a new concept is now defined. The *reliability self-information* (RSI), $I(R)$, is

$$\mathrm{I(R)} = -\log R, \tag{9–26}$$

while the *failure self-information* is

$$I(F) = -\log F, \tag{9–27}$$

where $F$ is the probability of failure, i.e.,

$$F = 1 - R. \tag{9–28}$$

The base of the logarithm is arbitrary, since different bases will simply add a constant bias. Since these quantities are measures of self-information, they enjoy many of the same properties that entropies do. For instance, multiplication can be replaced by addition, etc. The information-theoretic setting makes RSI easy to interpret in terms of Intelligent Control as proposed by Saridis and Valavanis (1988). The flow of knowledge resulting from the reliability self-information complies with the principle of Increasing Precision with Decreasing Intelligence (Saridis, 1989).

Once the reliabilities $R_{ik}$ have been found, it is a simple matter to find the corresponding RSI's, $I(R_{ik})$ from (9–26). It is also a simple matter to find the RSI of the entire subset if its elements are in a series relationship. The reliability of a series system (assuming independence) is (Ang and Tang, 1984)

$$R_s = R_1 R_2 \cdots R_n, \tag{9–29}$$

where $R_i$ denotes the reliability of the $i$th component. The RSI for a potential subset of algorithms $A_{\text{feas}_i}$ corresponding to $n$ specifications $(s_k)$ in a series relationship is found by combining (9–29) and (9–26):

$$I(R_i) = \sum_{k=1}^{n} I(R_{ik}). \tag{9–30}$$

Thus multiplication is replaced by addition when using RSI terms. On the other hand, if the set of specifications, $S_D$, contains parallel relationships, the parallel relationships must be simplified to one series term before (9–30) can be used. The reliability of a parallel arrangement (assuming independence) is

$$R_s = 1 - \prod_i (1 - R_i). \tag{9–31}$$

Rewriting (9–31) in terms of failure probabilities, we find

$$F = \prod_i F_i. \tag{9–32}$$

The total failure self-information of a parallel connection can be found from (9–27) and (9–32)

$$I(F) = \sum_i I(F_i). \tag{9–33}$$

As before, addition replaces multiplication. The RSI can be found from the failure self-information by the simple formula

$$I(R) = -\log(1 - \exp[-I(F)]). \tag{9–34}$$

The RSI found from (9–34) and (9–33) is the equivalent RSI of the parallel set of

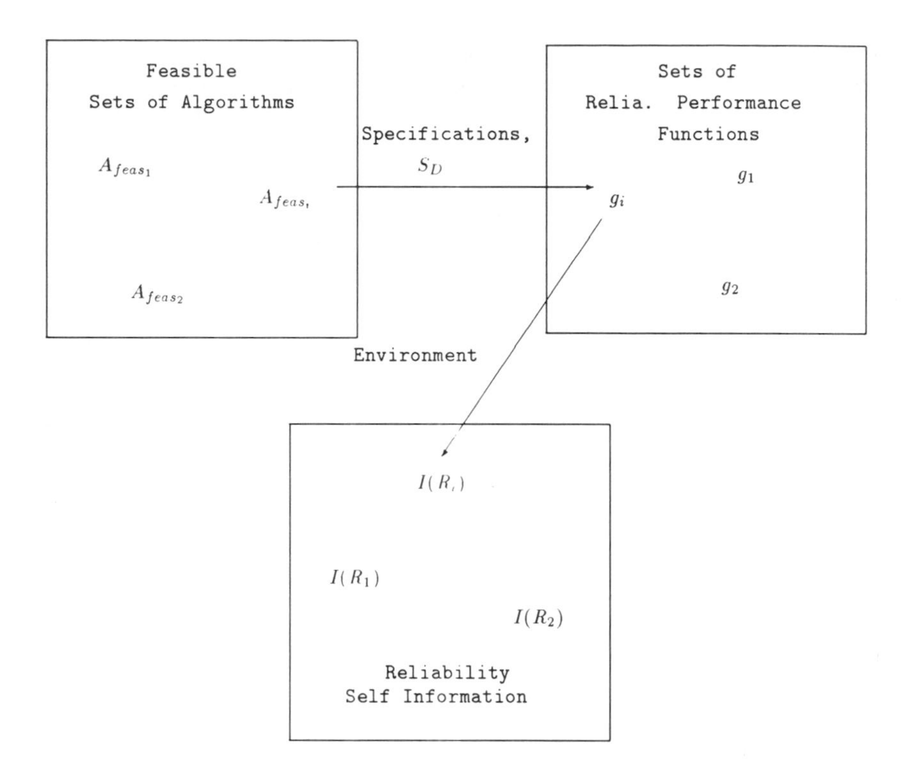


**FIGURE 9–3.** The set of desired specifications operates on the feasible algorithms to produce a set of reliability performance functions. The environment operates on the performance functions to produce a reliability self information.

specifications, so it can be combined with specifications in series with it using (9–30).

Naturally, once the RSI, $I(R_i)$, is found for each subset of feasible algorithms, $A_{feas_i}$, the performance of the alternate plans can easily be compared. The smaller the RSI, the more reliable is the plan. For a qualitative description of the reliability analysis procedure, see Figure 9–3.

To clarify these concepts further, the next section performs this analysis in detail for a robotic problem of practical significance.

## 4. RELIABILITY OF VISUAL POSITIONING

Visual positioning is an extremely important and highly developed facet of robotics. However, surprisingly few attempts have been made at analyzing the reliability of visual positioning. Thus, this section has a twofold purpose. First, it illustrates the ideas of the preceding sections so that the notation and concepts are made clear. Second, it provides a general reliability solution, useful for many hand/eye coordination tasks of a robotic manipulator.

> *Problem to Be Solved:* A manipulator has already moved close to a desired gripping point on an object using *a priori* knowledge. Owing to environmental uncertainties, the final movement cannot be completed without measuring the current position of the gripping point. Using a stereo vision system mounted on a separate arm, it is possible to view the object from $N$ different preprogrammed positions. Five inverse kinematics routines are available, and a library of $M$ computed torque compensators can be used for manipulator control. It is desired to make the final movement subject to two specifications—the total execution time and final position error are constrained with bounds $t_f$ and $e_{p_{max}}$, respectively. Find reliable subsets of control and sensing that are capable of satisfying these specifications.

The first step in the analysis is the definition of the desired set of specifications, $S_D$. For the given problem, the specifications are:

$s_1$ = {total execution time is less than a desired time, $t_f$}
$s_2$ = {final positioning error in the $l$th direction is less than a positioning threshold, $e_{p_{max_l}}$}

Next, the plans must be formed. In this case, each plan will consist of the triplet $A_i = \{V_i, K_i, G_i\}$, where $V_i$, $K_i$, and $G_i$ are the vision algorithm, inverse kinematics, and compensator used by the $i$th plan, respectively. Each algorithm is now examined in depth to find expressions for the statistical performance of each plan.

## 4–1. Analysis of the Vision System

As previously stated, the vision system consists of a pair of cameras mounted on the end of a separate robot arm. Although the stereo vision system may be positioned at $N$ different viewpoints to reduce problems with noise, assume there is only sufficient time to take measurements from one of the viewpoints. Each viewpoint will be regarded as a separate vision algorithm, thus the number of potential vision algorithms is also $N$. The accuracy in measuring the object's position will depend on the position of the object in the camera coordinate frame. Consequently, each viewpoint will have differing measurement statistics. These statistics can be found using the recently developed results of Lee and Kay (1990).

Lee and Kay first obtain linearized expressions for the position measurement error in the camera's frame caused by the image noise. Rewriting these expressions in terms of the object's measured position [similar to the approach taken in Rodriguez and Aggarwal (1990)] yields

$$e_{p_c} = \begin{bmatrix} e_x \\ e_y \\ e_z \end{bmatrix} = \begin{bmatrix} z_c/f & -z_c x_c/fB & z_c x_c/fB \\ 0 & z_c(B/2 - y_c)/fB & z_c(B/2 + y_c)/fB \\ 0 & -z_c^2/fB & z_c^2/fB \end{bmatrix} \begin{bmatrix} n_x \\ n_{yl} \\ n_{yr} \end{bmatrix} = M_c n_c \qquad (9\text{–}35)$$

where $e_x$, $e_y$ and $e_z$ = the position measurement errors in the camera coordinate frame
$x_c$, $y_c$, and $z_c$ = the measured object position in the camera frame
$f$ = the focal length
$B$ = the baseline
$n_x$, $n_{yl}$, and $n_{yr}$ = the image position errors.

Thus (9–35) can be used to find the position measurement error in the camera frame from the statistics of the image noise. The linearized orientation error is found by combining the noise from four measured points, which form an orthogonal coordinate frame. The linearized expression is

$$\delta = (M^T M)^{-1} M^T M' F n, \qquad (9\text{–}36)$$

where $T$ = the matrix transpose operation
$\delta$ = the orientation error in the camera frame
$M$ = a matrix formed from the measured camera coordinate frame positions of the four points
$M'$ = a constant matrix
$F$ = a matrix formed from the camera parameters and the measured camera frame positions of the four points
$n$ = a vector containing the image position errors of the four points

A vector containing both the position and orientation error due to image noise can then be formed

$$e_c = [e_{p_c}^T \delta^T]^T = Ln, \qquad (9\text{–}37)$$

where $L$ is determined by camera parameters and the four measured camera frame positions of the points.

The statistics of the image noise, $n$, are largely determined by the discretization in each pixel. Rodriguez and Aggarwal (1990) model the noise due to individual pixel errors as uniformly distributed. For purposes of noise rejection, assume that feature matching of centroids is used by the vision system. Since the centroid consists of an additive aggregation of the individual pixels, its distribution tends toward a Gaussian by the central limit theorem, so $n$ will be modeled as Gaussian, in accordance with Lee's model. Moreover, the elements of $n$ represent the $x$ and $y$ levels of image noise which are assumed to be independent based on the physical design of the cameras. Since the noise is zero mean, $n \sim N(0,C_v)$, where $C_v$ is a diagonal matrix.

Once the image measurements have been made, then $L$ may be computed, and the error statistics $e_c$ can be found. This information serves as an excellent performance check after the camera is positioned and measurements have been made. Unfortunately, it does not provide a method of evaluating each viewpoint without doing the time and computationally intensive repositioning of cameras and matching of features. Fortunately, an estimate of $e_c$ can be obtained without making the image measurements based on *a priori* knowledge. As the problem statement notes, an *a priori* estimate of the object's position/orientation in the base frame is known. This frame may be mapped into the $i$th camera coordinate system using the homogeneous coordinate transformation $T_b^{c_i}$, which maps points from the base to the $i$th camera coordinate system. Because the camera viewpoints are prespecified, the $T_b^{c_i}$ transformations can be calculated off-line and stored for later use at little cost to the system. Hence, the *a priori* estimate of the four points used for matching can be transformed from the base frame to each camera frame quite easily. Estimates of the error transformation for the $i$th viewpoint, denoted as $\hat{L}_i$, can then be calculated. The quality of $\hat{L}_i$ depends on the relative error of the *a priori* points in the camera frame. Since the cameras are kept at a respectable distance from the object to avoid collisions and interference with the manipulation task, the relative error is small, so initial analysis using $\hat{L}_i$ should be quite accurate. If the cameras are used in an extremely close proximity to the object, then $\hat{L}_i$ could conceivably be conditioned on the probability density function of the *a priori* estimates (Rodriguez and Aggarwal, 1990) to improve the estimate.

The time that each vision algorithm consumes consists of both a stochastic and a deterministic component. First, the cameras must be repositioned to the $i$th viewpoint. This is a preprogrammed movement, so it takes a deterministic quantity of time. Next, the measurement of the points must be made. The measurement time is random because of the variable time that the matching takes. Suppose that the total vision time for the $i$th algorithm (which includes the camera positioning time, the image processing time, and the time required for transformation to the base frame) is

$$t_{vi} \sim N(\mu_{t_{vi}}, \sigma_{t_{vi}}^2). \tag{9–38}$$

Once the statistics for positioning are found from (9–37) and those of timing are found from (9–38), the necessary information for the vision reliability analysis is complete, and the inverse kinematics analysis can begin.

## 4–2. Analysis of the Inverse Kinematics Routine

In order to act upon the vision system's perception, it is necessary to use an inverse kinematics routine to calculate the joint position corresponding to the Cartesian object position. (From this point forward, *position* will usually refer to the combined position/orientation.) This step is required because a joint-level-computed torque algorithm is used for servoing. Several inverse kinematics algorithms have been developed, and the choice of which algorithm to use depends on the time available and the accuracy required.

For this example, five inverse kinematics routines will be considered. The first two algorithms are proposed by Zhang and Paul (1990a, 1990b) for rapid solution using parallel processing. In order to produce a technique amenable to parallel computation, they note that the serial calculations can be parallelized approximately into six independent processes, with each processor finding a separate joint angle. The technique is approximate because the calculation of each joint angle depends on the previous joint angles in the kinematic chain. To decouple the calculations, the current joint position, rather than the desired joint position, is used in the inverse kinematics calculation. This algorithm will be denoted as $K_{\|1}$. As a later refinement, they suggest (Zhang and Paul, 1990a) using a linear extrapolation that improves the accuracy of the calculations while maintaining a parallel solution. Although parallelism is preserved, the linear extrapolation method, denoted as $K_{\|2}$, does require significantly more calculations (and hence time). Using a large number of measurements over the workspace of the robot, Zhang and Paul model the errors in the base frame statistically. Assuming that the sample mean and standard deviation of the error is measured, the most pessimistic distribution is Gaussian by the MEM, so the errors will be assumed to come from a Gaussian distribution. Note that the accuracy of these algorithms will depend on the position increment taking place from current to desired joint angles, so the measurements must be taken with a position increment of similar size to that required for the final movement. The computer operations required to calculate the inverse kinematics using either of these routines is fixed, so the time necessary is a deterministic constant for both $K_{\|1}$ and $K_{\|2}$.

The third method is probably the most commonly used inverse kinematics technique—simply use the nominal link parameters provided by the manufacturer in the standard serial calculations such as presented in Fu et al. (1987). This algorithm, denoted by $K_N$, offers greater accuracy than $K_{\|1}$ or $K_{\|2}$ at the cost of increased computation time because it must be processed on one serial CPU. However, the nominal link parameters are not exact due to manufacturing spreads, so error is still introduced by the using these parameters in the inverse kinematics computation. Since many commercially available manipulators are designed explicitly to have closed solution inverse kinematics for the nominal link parameters, it will be assumed that the nominal manipulator possesses a closed-form inverse kinematics solution. Unfortunately, if a calibration technique such as that proposed by Hayati et al. (1988) is employed, the resulting calibrated link parameters may not have closed-solution inverse kinematics. Consequently, to avoid the time-consuming iterative schemes for finding the inverse kinematics, nominal kinematics may be used even if calibrated link parameters are available. To measure the positioning accuracy using $K_N$, the errors are again measured over a large number of points in the workspace, and the maximum entropy Gaussian distribution with the sample mean and variance is applied. For a more refined analysis, it is possible to estimate the link parameters as described in Hayati et al. (1988). Next, through maximum likelihood estimation the positioning error (and its distribution) as a function of joint position is estimated (McInroy and Saridis, 1990b). In either case, since the nominal kinematics have a closed-form solution, the time required to perform the computation is fixed.

The fourth method, proposed by Hayati et al. (1988) and denoted as $K_J$, approximately updates the link parameters without iterative schemes by using the inverse manipulator Jacobian. Since the Jacobian is employed, the method is ill-conditioned in the neighborhoods of Jacobian singular points. Using singular value properties and maximum likelihood estimation, an asymptotically normal approximate upper bound on the normed positioning error can be found (McInroy and Saridis, 1990b). The bound can be used as a conservative estimate of the positioning accuracy. Although this statistical model does capture singularity effects quite nicely, only statistics for the norm (and not individual elements) are derived. Since several inverse kinematics algorithms are available, the refined estimate is not needed because the other algorithms can be employed when in the vicinity of singular points. As a result, accuracy statistics obtained by the sample mean and variance (as above) will be used for the majority of this case study. As in the previous algorithms, execution time is fixed.

The final algorithm, denoted as $K_U$, makes use of a calibration technique to find the updated link parameters for the particular manipulator under consideration. As alluded to previously, the updated kinematics may fail to possess a closed-form solution, so an iterative technique such as the Newton–Raphson algorithm or the Jacobi iterative method will be utilized [see Stone (1986) for details]. Again, this algorithm offers a trade-off: increased accuracy for increased time. As before, the errors are measured, and a Gaussian is used to model the error distribution. In contrast, the time for executing $K_U$ depends on the number of iterations required. As a result, the execution time when using the iterative technique is stochastic. To model the statistics, the sample mean and variance of the execution time are found, and the most pessimistic Gaussian distribution is assumed as a conservative measure.

To summarize, five inverse kinematics algorithms are available that play time against accuracy. The algorithms are:

$K_{\|1}$ = {Zhang and Paul's parallel inverse kinematics}
$K_{\|2}$ = {Zhang and Paul's parallel inverse kinematics with linear extrapolation}
$K_N$ = {Serial inverse kinematics using nominal link parameters}
$K_J$ = {Update kinematics using the manipulator Jacobian}
$K_U$ = {Iterative inverse kinematics using updated link parameters}

The choice of inverse kinematics routine depends on the set of specifications, $S_D$. The specifications, in turn, interact with the statistical parameters of the algorithm. These statistical parameters will be denoted as

$$e_{ki} \sim N(0, C_{ki}), \qquad (9\text{–}39)$$

$$t_{ki} \sim N(\mu_{t_{ki}}, \sigma^2_{t_{ki}}), \qquad (9\text{–}40)$$

where $e_{ki}$ is the base frame Cartesian positioning error attributed to the $i$th inverse kinematics algorithm, and $t_{ki}$ is the time required for the inverse kinematics.

## 4–3. Analysis of the Control System

To control the movements of the manipulator, a computed torque control law is used. This control law is based on the Euler–Lagrange dynamic equation:

$$\tau = D(\Theta)\ddot{\Theta} + NL(\Theta, \dot{\Theta}), \qquad (9\text{–}41)$$

where $\tau$ = the torque applied by the motors
$\Theta$ = the joint position vector
$\dot{\Theta}$ = the joint velocity vector
$\ddot{\Theta}$ = the joint acceleration vector
$D(\Theta)$ = the inertia matrix
$NL(\Theta, \dot{\Theta})$ = a vector of torques due to Coriolis, gravity, centripetal, and friction nonlinearities

An effective method of controlling an arm with dynamics of the form (9–41) is the computed torque technique described in Saridis (1985), where the control function is defined as

$$\tau = D(\Theta)E + NL(\Theta, \dot{\Theta}), \qquad (9\text{–}42)$$

where $E$ is a compensation term. Since $D(\Theta)$ is invertible due to the physical laws of inertia, this implies that

$$s^2\Theta = E(s) = G(s)(\Theta_d(s) - \Theta(s)), \tag{9–43}$$

where $s$ denotes Laplace transformation, $G(s)$ is a compensator, and $\Theta_d$ is the desired position. The multivariable transfer function for the system is

$$\Theta = (1/s^2)[I + (1/s^2)G(s)]^{-1}G(s)\Theta_d. \tag{9–44}$$

If $G(s)$ is a diagonal matrix, then all joints are decoupled, and they can therefore be treated separately. Since this decoupling significantly decreases the complexity of the design, assume that all of the available compensators are diagonal. Furthermore, assume that the compensators have been designed so that the characteristic equation

$$\theta_j = \frac{\omega_{n_{ij}}^2 \theta_{d_j}}{s^2 + 2\zeta_{ij}\omega_{n_{ij}}s + \omega_{n_{ij}}^2} \tag{9–45}$$

describes the response of the $j$th joint to the $i$th compensator. Equation (9–45) can be used if (Dorf, 1989):

1. There are no zeros in the transfer function near the dominant second order poles.
2. The real part of the second-order roots is less than 1/10th the real part of higher-order roots.

If these criteria are met, then the settling time and position error due to a step in position can be found in a straightforward manner.

The settling time of the $j$th joint when using the $i$th compensator is

$$t_{G_{ij}} = \frac{4}{\zeta_{ij}\omega_{n_{ij}}}. \tag{9–46}$$

Since the control system does not allow steady-state discrepancies between the commanded joint position and the actual point position for step inputs, the steady-state position error is due to the sensing estimation and measurement errors as well as the inverse kinematic transformation errors. As a result, the position error for the $i$th plan is

$$e_{p_i} = x_{\text{obj}} - \hat{x}_i + e_{ki}, \tag{9–47}$$

where $x_{obj}$ is the base frame position/orientation vector of the object and $\hat{x}_i$ is the measured position/orientation found by the sensing system. Since the vision system is mounted on a separate robot, $\hat{x}_i$ contains errors due to both the image noise present in the vision system and the forward kinematic transformation errors used to transform the camera measurements to the base frame. The errors due to image noise have been developed in Section 4–1. The forward kinematic errors can be found by utilizing concepts similar to those found in Section 4–2. Assuming that the vision system arm has been previously calibrated using Hayati et al.'s (1988) technique, a large number of measurements have been made concerning the positional accuracy. Since forward kinematics can easily use the updated link parameters without any need of iterative techniques such as the inverse kinematics require, the updated parameters are used in the forward kinematics. Hayati et al.'s technique finds a least-squares solution for these updated parameters based on the error in the end-effector's coordinate frame of reference. Use of the updated parameters found via this least-squares solution then produces a positioning error with zero mean and covariance equal to the residuals of the solution, $C_{FK}$. Lacking any other information, the most conservative distribution is Gaussian.

The statistics for the settling time and position error have been found; therefore, it is now possible to combine the information to find the feasible plans.

## 4–4. Selection of Feasible Plans

Because the total number of possible plans for accomplishing the task is $5NM$, a detailed analysis of every possibility can be very time consuming. To avoid a complete analysis for all possibilities, a subset of the plans (the feasible plans) will be found that satisfies necessary, but not sufficient, conditions for reliable operation as described in Section 2.2.

Consider the individual tasks that must be performed in order to solve the given problem. First, the cameras must be moved to one of the $N$ viewpoints. From the viewpoint, image processing is used to measure the object's position in the camera coordinate system. The measurement will, of course, be imperfect due to image noise. In addition, it will take a variable length of time due to the use of stereo matching. The measured position is then transformed to the base frame. The transformation depends on the forward kinematic accuracy of the camera manipulator, so again errors are introduced. Based on the transformed measurement, an inverse kinematics algorithm is selected to find a joint position for servoing the manipulator toward. The inverse kinematics introduces stochastic position errors and (for $K_U$) requires a stochastic period of time. Finally, the joint servoing of the motors using the computed torque technique is performed.

The first specification, $s_1$, constrains execution time. The time specification must be satisfied by all of the joints since the joints are servoing simultaneously. The total execution time for the $j$th joint while using the $i$th plan is

$$t_{ij} = t_{vi} + t_{ki} + t_{G_{ij}}. \tag{9–48}$$

Assuming independence, (9–38), (9–40), and (9–46) yield

$$E\{t_{ij}\} = \mu_{t_{ij}} = \mu_{t_{vi}} + \mu_{t_{ki}} + \frac{4}{\zeta_{ij}\omega_{n_{ij}}}, \tag{9–49}$$

$$\mathrm{Cov}(t_{ij}) = \sigma^2_{t_{vi}} + \sigma^2_{t_{ki}}. \tag{9–50}$$

To produce an entropy specification with zero mean, the timing specification will be considered as bounds $(-[t_f - \mu_{t_{ij}}],[t_f - \mu_{t_{ij}}])$. The quantity $[t_f - \mu_{t_{ij}}]$ is positive for reliable systems, so the resulting entropy from Table 9–1 is $H(s_{1_{ij}}) = \ln(2[t_f - \mu_{t_{ij}}])$. The entropy of (9–50) is $H(A_{i_1}) = \ln\sqrt{2\pi e(\sigma^2_{t_{vi}} + \sigma^2_{t_{ki}})}$. As a result, the timing constraints will be met if

$$\ln\sqrt{2\pi e(\sigma^2_{t_{vi}} + \sigma^2_{t_{ki}})} < \ln(2[t_f - \mu_{t_{ij}}]) \tag{9–51}$$

is satisfied for all of the joints, $j$.

The second specification constrains final positioning error. For the $l$th direction it consists of bounds $(-e_{p_{\max_l}}, e_{p_{\max_l}})$. Since the specifications for each direction are independent, the total specification entropy from Table 9–1 is

$$H(s_2) = \sum_{l=1}^{6} \ln 2e_{p_{\max_l}}. \tag{9–52}$$

The joint positioning entropy is the sum of several independent entropies. First, an entropy arises due to the image noise, $H(v_{\text{image}_i})$. Because the image noise is assumed to be normally distributed (9–8) and Papoulis (1984) produce the entropy in the camera frame of

$$H(v_{\text{image}_i}) = \ln\sqrt{(2\pi e)^6 \det(C_v)} + E\{\ln|\det(\hat{L}_i)\}. \tag{9–53}$$

Note that the entropy in the camera frame of the oriented point can be determined (to a good approximation) without actually moving to the $i$th viewpoint and taking measurements. Moreover, since the entropy of an oriented point is invariant with respect to homogeneous coordinate frame transformations, the entropy does not need to be transformed to the base frame—there is no need of calculating the Jacobian for the transformation, etc.

Next, the entropy due to camera forward kinematic positioning errors, $H(v_{FK})$, must be accounted for. From the calibration discussed in Section 4–3

$$H(v_{FK}) = \ln \sqrt{(2\pi e)^6 \det(C_{FK})}. \tag{9–54}$$

Although the calibration, and hence $C_{FK}$, are with respect to the end-effector's frame of reference, the entropy invariance property implies that $H(v_{FK})$ will remain unchanged in any homogeneous coordinate system.

Finally, the entropy introduced by the inverse kinematics routine must be found. From (9–39), it is

$$H(k_i) = \ln \sqrt{(2\pi e)^6 \det(C_{ki})}. \tag{9–55}$$

Because each of these sources of error is independent, the total entropy is equal to the sum, i.e.,

$$H(A_{i_2}) = H(v_{\text{image}_i}) + H(v_{FK}) + H(k_i). \tag{9–56}$$

The positioning constraints will then be met by the $i$th plan if

$$H(A_{i_2}) \leq H(s_2). \tag{9–57}$$

The feasible plans can quickly be found by ensuring that (9–51) and (9–57) are both satisfied. For instance, the timing constraints can be used to immediately remove those subsets that cannot finish in the desired final time, $t_f$. The positioning constraints can then be used to further the elimination of unsuitable plans. Since the positioning analysis is performed on all six degrees of freedom jointly, it is conceivable that a plan may satisfy the positioning constraint by obtaining extremely small error in one direction, while the specification is actually exceeded in another direction. Judicious choice of specifications will usually render this point moot. This means that appropriate selection of specifications can often be used to eliminate explicit calculation of reliabilities for the feasible plans. However, if a desired level of reliability must be satisfied, then it is necessary to perform a reliability analysis of the feasible plans.

## 4–5. Calculation of Reliability

The entropy constraints generated in the last section implicitly ensure a level of reliability. In order to explicitly find the reliability for a particular plan, further analysis is warranted.

First, the specifications, $S_D$, are used in conjunction with the mathematical laws governing the system to produce reliability performance functions corresponding to each algorithm. The timing RPF for the $i$th plan on the $j$th joint follows in a straightforward manner:

$$g_{ij1} = t_f - t_{ij}. \tag{9–58}$$

Since $t_{ij}$ consists of a linear combinations of times, by (9–48) the reliability index corresponding to $g_{ij1}$ can be found from (9–20). The first reliability index for the $A_{\text{feas}_i}$ set of algorithms on the $j$th joint is thus

$$\beta_{\text{ij1}} = \frac{t_f - \mu_{t_{vi}} - \mu_{t_{ki}} - 4/(\zeta_{ij}\omega_{n_{ij}})}{\sqrt{(\sigma^2_{t_{vi}} + \sigma^2_{t_{ki}})}} . \tag{9–59}$$

By using a look-up table for the zero mean normal cumulative distribution function, the reliability can easily be found from (9–21). Moreover, the RSI is found from (9–26):

$$I(R_{ij1}) = -\log \Phi\ [\beta_{ij1}]. \tag{9–60}$$

Note that all of the reliability self-information terms for execution time can be determined off-line. This is quite a time-saving feature, because it allows many of the unreliable plans [i.e., plans with large $I(R_{ij1})$ terms] to be identified before execution begins.

Since the positioning errors have been modeled as Gaussian, the reliability can be calculated from the Gaussian standard normal distribution function. Following an approach similar to Lee and Kay's (1990), the first-order error in positioning (not including orientation) of the $i$th feasible plan is

$$e_i = R_i^b M_{ci} n_c + M_{FK_i}\delta_{FK} + e_{FK} + e_{p_{ki}}, \tag{9–61}$$

where $R_i^b$ = the rotation matrix from the $i$th viewpoint to the base frame
$M_{ci}$ and $n_c$ are given by (9–35) with $i$ representing the viewpoint

$$M_{FK_i} = \begin{bmatrix} 0 & -n_z x_c - s_z y_c + a_z z_c & -n_y x_c - s_y y_c - a_y z_c \\ -n_z x_c - s_z y_c - a_z z_c & 0 & n_x x_c + s_x y_c + a_x z_c \\ n_y x_c + s_y y_c + a_y z_c & -n_x x_c - s_x y_c - a_x z_c & 0 \end{bmatrix}$$

$\delta_{FK}$ = the orientation error due to the forward kinematics of the camera positioning system

$e_{FK}$ = the position error due to approximate forward kinematics of the camera's arm

$e_{p_{ki}}$ = the positioning error caused by approximate inverse kinematics for the manipulator

Since all of the random vectors ($n_1$, $\delta_{FK}$, $e_{FK}$, and $e_{p_{ki}}$) are zero mean and independent, the expected value of $e_i$ is also zero. The covariance matrix of $e_i$ is

$$\mathrm{Cov}(e_i) = R_i^b M_{ci}\,\mathrm{Cov}(n_1)(R_i^b M_{ci})^T + M_{FK_i}\,\mathrm{Cov}(\delta_{FK})M_{FK_i}^T + \mathrm{Cov}(e_{FK}) + \mathrm{Cov}(e_{p_{ki}}). \tag{9–62}$$

The reliability of the $i$th plan in meeting the $l$th positioning specification is then

$$R_{il2} = P\{-e_{p_{\max_l}} \le e_{p_{i_l}} \le e_{p_{\max_l}}\} = 2\Phi(e_{p_{\max_l}}/\sigma_{e_{i_l}}) - 1, \qquad l = 1, \ldots, 3, \tag{9–63}$$

where $\sigma^2_{e_{i_l}}$ is the $(l,l)$th element of $\mathrm{Cov}(e_i)$.

The reliability in orienting is found in a similar manner. From Lee and Kay's paper and (9–36) the orientation error in the base frame is

$$\delta_i = N_{ci}(M_i^T M_i)^{-1} M_i^T M' F_i n + \delta_{FK} + \delta_{ki} = L_i^b n + \delta_{FK} + \delta_{ki}, \tag{9–64}$$

where $N_{ci}$ is determined by the column vectors of the camera to base transformation ($R_i^{b-1} = [n\ s\ a]$),

$$N_{ci} = \begin{bmatrix} s_y a_z - a_y s_z & -s_x a_z + a_x s_z & a_y s_x - s_y a_x \\ n_z a_y - a_z n_y & -n_z a_x + a_z n_x & n_y a_x - a_y n_x \\ s_z n_y - n_z s_y & -n_x s_z + s_x n_z & n_x s_y - s_x n_y \end{bmatrix}, \tag{9–65}$$

and $\delta_{ki}$ is the orientation error caused by the inverse kinematics inaccuracies. Similar to the position error, the orientation error is zero mean with covariance

$$\mathrm{Cov}(\delta_i) = L_i^b \mathrm{Cov}(n)(L_i^b)^T + \mathrm{Cov}(\delta_{FK}) + \mathrm{Cov}(\delta_{ki}), \tag{9–66}$$

The reliability of the $i$th plan in meeting the second specification in the $l$th orientation is then

$$R_{il2} = 2\Phi(e_{p_{\max_l}}/\sigma_{\delta_{il}}) - 1, \tag{9–67}$$

where $\sigma^2_{\delta_{il}}$ is the $(l,l)$th element of $\mathrm{Cov}(\delta_i)$. The RSI terms follow immediately

$$I(R_{il2}) = -\log(R_{il2}). \tag{9–68}$$

Equations (9–60) and (9–68) are explicit functions relating the $i$th feasible subset of algorithms, $A_{\mathrm{feas}_i}$, and the environment and design stochastic variables to the RSI. The equations must now be combined to form one total RSI for each subset of algorithms.

Since all of the specifications must be satisfied to ensure that the desired standards are met and the correlation between terms (reflected in the off-diagonal terms of the covariance matrices) is generally minimal, the system forms a series connection. (If terms cannot be considered independent, then the full correlation matrix must be used.) To find the RSI of a particular specification, the RSI terms found from Eqs. (9–60) and (9–68) are summed for all six directions:

$$I(R_{i1}) = \sum_{j=1}^{6} I(R_{\mathrm{i}l1}), \tag{9–69}$$

$$I(R_{i2}) = \sum_{l=1}^{6} I(R_{\mathrm{i}l2}), \tag{9–70}$$

The total RSI for a particular subset of algorithms, $A_{\mathrm{feas}_i}$, is found from (9–30), (9–69), and (9–70) to be the sum of the RSI's for each specification, i.e.,

$$I(R_i) = I(R_{i1}) + I(R_{i2}). \tag{9–71}$$

A threshold can be set on $I(R_i)$ to discriminate reliable from unreliable plans. Those plans with a total reliability self-information less than the threshold will then be considered reliable. In this manner, a specified level of reliability in meeting the desired requirements can be achieved.

## 4–6. Simulation Results

Although the complete case study has not been simulated to date, two of the inverse kinematic routines ($K_N$ and $K_J$) have been simulated using Puma 560 kinematics. The positioning statistics are derived as a function of the joint position as detailed in McInroy and Saridis (1990c). Consequently, the most appropriate positioning algorithm for a given specification depends on the joint angle. Figure 9–4 plots the actual positioning accuracy of the two algorithms at a number of joint positions. As expected, the Jacobian update scheme ($K_J$) demonstrates considerable accuracy at some locations in the workspace, but lacks precision near Jacobian singularities. The reliability of the two algorithms for the same set of joint positions is shown in Figure 9–5. Since the positioning accuracy of each algorithm varies throughout joint space, the algorithm of choice also varies. For instance, at point 4, $K_N$ is preferable, while $K_J$ yields higher reliability for point 6. By deriving this information, the reliability concepts developed herein facilitate the analytic design of Intelligent Machines.

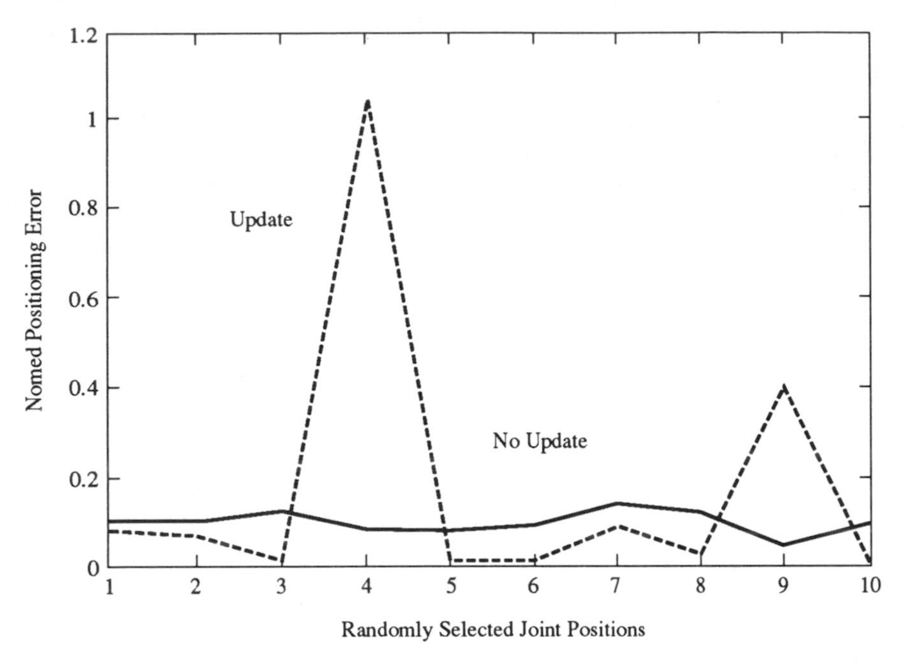


**FIGURE 9–4.** The normed positioning error without update ($K_N$), and with the update ($K_J$). Note that the 4th and 9th updates produced larger position errors than nominal kinematics.

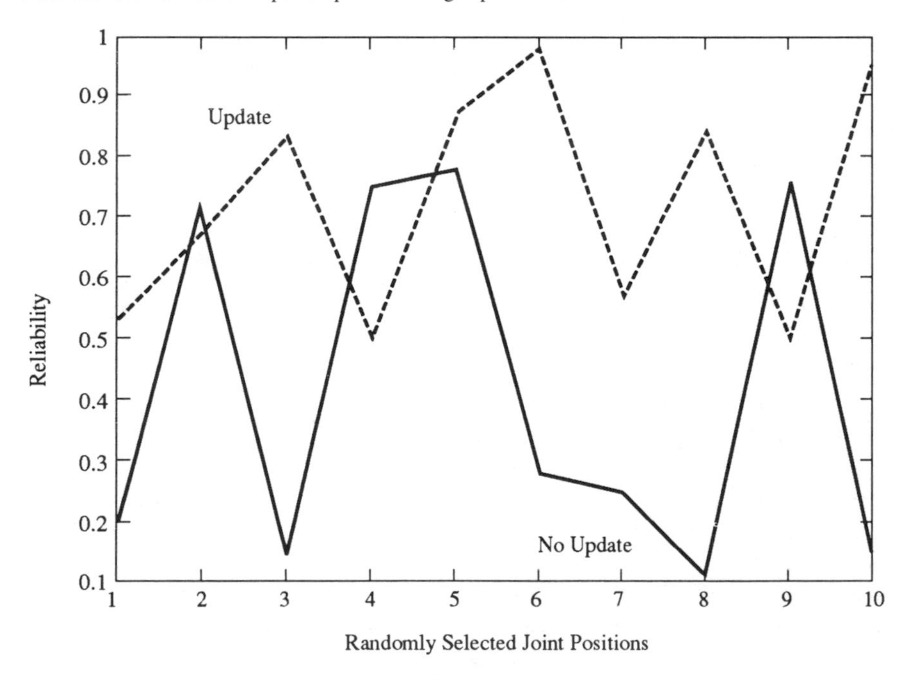


**FIGURE 9–5.** The positioning reliability without update ($K_N$), and with the update ($K_J$). The positioning specification is $e_{p\ \max} = 0.1$.

## 5. CONCLUSIONS

Concepts from reliability and information theory have been brought together and utilized for selecting reliable plans to execute a given task. The approach developed holds several advantages. First, it treats the topics of reliability, control, and sensing in a unified, analytic manner. Second, entropy is invariant with respect to coordinate frame transformations, so distributed systems are easily accommodated. Third, the method is not restricted to purely Gaussian distributions. Finally, the entropy formulation provides a consistent representation throughout all levels in a hierarchically Intelligent Machine. As the example illustrates, the technique is applicable to problems of practical significance.

## 6. ACKNOWLEDGMENTS

This work was completed under NASA grant NGT-50357, and it was also partially supported by NASA grant NAGW-1333.

**References**

Ang, A. H., Tang, W. H. 1984. *Probability Concepts in Engineering Planning and Design. Volume II—Decision, Risk and Reliability*. New York: Wiley.

Azadivar, Farhad. 1987. The effect of joint positions errors of industrial robots on their performance in manufacturing operations. *IEEE Journal of Robotics and Automation* RA-3(2):109–114.

Blostein, Steven D. and Huang, Thomas S. 1987. Error analysis in stereo determination of 3-d point positions. *IEEE Transactions on Pattern Analysis and Machine Intelligence* PAMI-9(6):752–765.

Brooks, R. A. 1982. Symbolic error analysis and robot planning. *The International Journal of Robotics Research* 1(4):29–68.

Dorf, R. C. 1989. *Modern Control Systems*. 5th ed. Reading, MA: Addison-Wesley.

Durrant-Whyte, Hugh F. 1988. *Integration, Coordination and Control of Multi-Sensor Robot Systems*. Boston, MA: Kluwer Academic Publishers.

Fu, K. S., Gonzalez, R. C., and Lee, C. S. G. 1989. *Robotics: Control, Sensing, Vision, and Intelligence*. New York: McGraw-Hill.

Harr, M. E. 1987. *Reliability-Based Design in Civil Engineering*. NY: McGraw-Hill.

Havel, I. M. and Kramosil, I. 1978. A stochastic approach to robot plan formation. *Kybernetika* 14(3):143–173.

Hayati, S., Tso, K., and Roston, G. 1988. Robot geometry calibration. In *IEEE International Conference on Robotics and Automation,* Volume 2, pp. 947–951.

Jaynes, Edwin T. 1957. Information theory and statistical mechanics. *Physical Review* 106(4):620–630.

Jaynes, Edwin T. 1982. On the rationale of maximum-entropy methods. *Proceedings of the IEEE* 70(9):939–952.

Koomen, C. J. 1985. The entropy of design: A study on the meaning of creativity. *IEEE Transactions on Systems, Man, and Cybernetics* SMC-15(1):16–30.

Kalata, Paul and Priemer, Roland. 1974. On minimal error entropy stochastic approximation. *International Journal of Systems Science* 5(9):895–906.

Kalata, Paul and Priemer, Roland. 1979. Linear prediction, filtering, and smoothing: An information-theoretic approach. *Information Sciences* 17:1–14.

Kyriakopoulos, Konstantinos. 1988. Private communication.

Lee, Sukhan and Kay, Youngchul. 1990. An accurate estimation of 3-d position and orientation of a moving object for robot stereo vision: Kalman Filter approach. In *IEEE International Conference on Robotics and Automation,* Volume 1, pp. 414–419.

McInroy, John E. and Saridis, George N. 1990a. Reliability analysis in intelligent machines. *IEEE Transactions on Systems, Man, and Cybernetics* 20(4):950–956.

McInroy, John E. and Saridis, George N. 1990b. Reliable high precision positioning for intelligent machines. In *5th IEEE Symposium on Intelligent Control,* Volume I.

Papoulis, Athanasios. 1984. *Probability, Random Variables, and Stochastic Processes.* 2nd ed. NY: McGraw-Hill.

Parkinson, David B. Computer solution for the reliability index. *Engineering Structures* 2:57–62.

Parkinson, D. B. 1982. The application of reliability methods to tolerancing. *ASME Journal of Mechanical Design* 104:612–618. July 1982.

Rodriguez, Jeffrey J. and Aggarwal, J. K. 1990. Stochastic analysis of stereo quantization error. *IEEE Transactions on Pattern Analysis and Machine Intelligence* PAMI-12(5):467–470.

Saridis, G. N. ed. 1985. *Advances in Automation and Robotics: Volume 1.* Greenwich, CT: JAI Press.

Saridis, George N. 1988. Entropy formulation of optimal and adaptive control. *IEEE Transactions on Automatic Control.* 33(8):713–721.

Saridis, George N. 1989. Analytic formulation of the principle of increasing precision with decreasing intelligence for intelligent machines. *Automatica* 25(3):461–67.

Smith, Randall C. and Cheeseman, Peter. 1986. On the representation and estimation of spatial uncertainty. *The International Journal of Robotics Research* 5(4):56–68.

Smith, R. E. and Gini, M. 1986. Reliable real-time robot operation employing intelligent forward recovery. *Journal of Robotic Systems* 3(3):281–300.

Shinozuka, M. 1983. Basic analysis of structural safety. *ASCE Journal of Structural Division* 3(109).

Stone, Henry Wallentin. 1986. *Kinematic Modeling, Identification, and Control of Robotic Manipulators.* PhD thesis, Carnegie Mellon University, Pittsburgh, PA.

Saridis, G. N. and Valavanis, K. P. 1988. Analytical design of intelligent machines. Automatica 24(2):123–33.

Taylor, P. M., Halleron, I., and Song, X. K. The application of a dynamic error framework to robotic assembly. In *IEEE International Conference on Robotics and Automation,* Volume I, pp. 170–175.

Taylor, G. E. and Taylor, P. M. 1988. Dynamic error probability vectors: A framework for sensory decision making. In *IEEE International Conference on Robotics and Automation.* Volume II, pp. 1096–1100.

Valavanis, K. P. and Saridis, G. N. 1988. Information-theoretic modeling of intelligent robotic systems. *IEEE Transactions on Systems, Man, and Cybernetics* 18(6):852–872.

Wilks, Samuel S. 1962. *Mathematical Statistics*. NY: Wiley.
Wang, Fei-Yue and Saridis, George N. 1989. The coordination of intelligent robots: A case study. In *IEEE International Symposium on Intelligent Control,* pp. 506–511.
Zacks, Shelemyahu. 1971. *The Theory of Statistical Inference*. NY: Wiley.
Zhang, Hong and Paul, Richard P. 1990a. A parallel inverse kinematics solution for robot manipulators based on multiprocessing and linear extrapolation. In *IEEE International Conference on Robotics and Automation*. Volume I, pp. 468–474.
Zhang, Hong and Paul, Richard P. 1990b. A parallel solution to robot inverse kinematics. In *IEEE International Conference on Robotics and Automation*. Volume II, pp. 1140–1145.

# Index

# Index